Springer Proceedings in Mathematics & Statistics

Volume 63

For further volumes:
http://www.springer.com/series/10533

Springer Proceedings in Mathematics & Statistics

This book series features volumes composed of select contributions from workshops and conferences in all areas of current research in mathematics and statistics, including OR and optimization. In addition to an overall evaluation of the interest, scientific quality, and timeliness of each proposal at the hands of the publisher, individual contributions are all refereed to the high quality standards of leading journals in the field. Thus, this series provides the research community with well-edited, authoritative reports on developments in the most exciting areas of mathematical and statistical research today.

Ettore Lanzarone • Francesca Ieva
Editors

The Contribution of Young Researchers to Bayesian Statistics

Proceedings of BAYSM2013

Editors
Ettore Lanzarone
CNR-IMATI
Milan, Italy

Francesca Ieva
Politecnico di Milano
Milan, Italy

ISSN 2194-1009   ISSN 2194-1017 (electronic)
ISBN 978-3-319-02083-9   ISBN 978-3-319-02084-6 (eBook)
DOI 10.1007/978-3-319-02084-6
Springer Cham Heidelberg New York Dordrecht London

Library of Congress Control Number: 2013954407

© Springer International Publishing Switzerland 2014
This work is subject to copyright. All rights are reserved by the Publisher, whether the whole or part of the material is concerned, specifically the rights of translation, reprinting, reuse of illustrations, recitation, broadcasting, reproduction on microfilms or in any other physical way, and transmission or information storage and retrieval, electronic adaptation, computer software, or by similar or dissimilar methodology now known or hereafter developed. Exempted from this legal reservation are brief excerpts in connection with reviews or scholarly analysis or material supplied specifically for the purpose of being entered and executed on a computer system, for exclusive use by the purchaser of the work. Duplication of this publication or parts thereof is permitted only under the provisions of the Copyright Law of the Publisher's location, in its current version, and permission for use must always be obtained from Springer. Permissions for use may be obtained through RightsLink at the Copyright Clearance Center. Violations are liable to prosecution under the respective Copyright Law.
The use of general descriptive names, registered names, trademarks, service marks, etc. in this publication does not imply, even in the absence of a specific statement, that such names are exempt from the relevant protective laws and regulations and therefore free for general use.
While the advice and information in this book are believed to be true and accurate at the date of publication, neither the authors nor the editors nor the publisher can accept any legal responsibility for any errors or omissions that may be made. The publisher makes no warranty, express or implied, with respect to the material contained herein.

Printed on acid-free paper

Springer is part of Springer Science+Business Media (www.springer.com)

Preface

This volume includes a selection from the contributions presented at the first Bayesian Young Statistician Meeting (BAYSM 2013). The conference was held at the Institute of Applied Mathematics and Information Technology (IMATI) of the National Research Council of Italy (CNR) in Milan, Italy, on June 5 and 6, 2013.

This conference provided an opportunity for M.S. students, Ph.D. students, postdoctoral scholars, and young researchers dealing with Bayesian statistics to connect with the Bayesian community at large, to exchange ideas, and to know people working in the same field for creating future networks. The aim was to create a scientific forum for the next generation of researchers in Bayesian statistics. In fact, the workshop encouraged discussion and promoted further research in all those fields where Bayesian statistics may be employed. The conference had about 100 participants from 20 different countries and 44 accepted contributions. The conference was divided into 6 talk sessions and 1 poster session, and it was opened by 2 keynote lectures. Finally, the presence of 7 senior discussants allowed participants to get advices and comments to their current researches. Thanks to the great success and interest of participants, BAYSM 2013 became the first of a series of conferences, devoted to young students and researchers dealing with Bayesian statistics. As an example, the second edition will hold in Vienna, Austria, in September 2014. Any other information can be found at the permanent website of the conference www.mi.imati.cnr.it/conferences/BAYSM2013.

This volume is structured in five parts, each one dealing with one of the mainstreams treated within the sessions: theoretical methods, computational methods, application of Bayesian statistics to real cases from industry and other applicative contexts, application to life sciences and healthcare, stochastic processes, and models for finance and economics. Finally, the sixth chapter is devoted to one of the keynote lectures, dealing with the publishing process of papers in statistics and, more in general, of scientific papers.

We wish to thank all of the authors who made the conference and this volume possible. We wish to Thanks are also due to reviewers, to senior discussants, and to keynote speakers for their help, their valued suggestions, and their fundamental contributions. Finally, we thank our colleagues at CNR-IMATI and Politecnico di Milano for supporting us in this exciting experience.

Milan, Italy Ettore Lanzarone
Milan, Italy Francesca Ieva

Contents

Part I Theoretical Bayes

1 **A Nonparametric Model for Stationary Time Series** 3
   Isadora Antoniano-Villalobos and Stephen G. Walker

2 **Estimation of Optimally Combined-Biomarker Accuracy in the Absence of a Gold Standard Reference Test** 7
   Leandro Garcia Barrado, Elisabeth Coart, and Tomasz Burzykowski

3 **On Bayesian Transformation Selection: Problem Formulation and Preliminary Results** 11
   E. Charitidou, D. Fouskakis, and I. Ntzoufras

4 **A Simple Proof for the Multinomial Version of the Representation Theorem** 15
   Marcio A. Diniz and Adriano Polpo

5 **A Sequential Monte Carlo Framework for Adaptive Bayesian Model Discrimination Designs Using Mutual Information** 19
   Christopher C. Drovandi, James M. McGree, and Anthony N. Pettitt

6 **Joint Parameter Estimation and Biomass Tracking in a Stochastic Predator–Prey System** 23
   Laura Martín-Fernández, Gianni Gilioli, Ettore Lanzarone, Joaquín Míguez, Sara Pasquali, Fabrizio Ruggeri, and Diego P. Ruiz

7 **Adaptive Bayes Test for Monotonicity** 29
   Jean-Bernard Salomond

8	Bayesian Inference on Individual-Based Models by Controlling the Random Inputs 35
	Michael Spence and Paul Blackwell
9	Consistency of Bayesian Nonparametric Hidden Markov Models 41
	Elodie Vernet
10	Bayesian Methodology in the Stochastic Event Reconstruction Problems 45
	Anna Wawrzynczak, Piotr Kopka, and Mieczyslaw Borysiewicz

Part II Computational Bayes

11	Efficient Fitting of Bayesian Regression Models with Spatio-Temporally Varying Coefficients 53
	Mark Bass and Sujit Sahu
12	PAWL-Forced Simulated Tempering 61
	Luke Bornn
13	Approximate Bayesian Computation for the Elimination of Nuisance Parameters 67
	Clara Grazian
14	Reweighting Schemes Based on Particle Methods 73
	Reinaldo Marques and Geir Storvik
15	A Bayesian Nonparametric Framework to Inference on Totals of Finite Populations 77
	Juan Carlos Martínez-Ovando, Sergio I. Olivares-Guzmán, and Adriana Roldán-Rodríguez
16	Parallel Slice Sampling 81
	Teresa Pietrabissa and Simone Rusconi
17	Approximate Bayesian Computation in Quantile Regression 85
	Antonio Pulcini

Part III Bayes @ Work: Appraisal of Applications to the Real World

18	Spatiotemporal Model for Short-Term Predictions of Air Pollution Data 91
	Francesca Bruno and Lucia Paci
19	Predicting Rainfall Fields from Lightning Records: A Hierarchical Bayesian Approach 95
	Edmondo Di Giuseppe, Giovanna Jona Lasinio, Massimiliano Pasqui, and Stanislao Esposito

Contents

20 Bayesian Approach to Environmental Problem Based on PFLOTRAN Package 101
Orest Dorosh, Henryk Wojciechowicz, and Piotr Kopka

21 Bayesian Hierarchical Modeling of Growth via Gompertz Model: An Application in Poultry 105
Emre Karaman, Ebru Kaya, Dogan Narinc, and Mehmet Z. Firat

22 Bayesian Prediction of SMART Power Semiconductor Lifetime with Bayesian Networks 109
Kathrin Plankensteiner, Olivia Bluder, and Jürgen Pilz

23 Consumer-Oriented New-Product Development in Fruit Flavor Breeding: A Bayesian Approach 113
Lebeyesus M. Tesfaye, Ivo A. van der Lans, Marco C.A.M. Bink, Bart Gremmen, and Hans C.M. van Trijp

24 Bayesian Layer Counting in Ice-Cores: Reconstructing the Time Scale 121
J.J. Wheatley, P.G. Blackwell, N.J. Abram, and E.W. Wolff

Part IV A Bayesian Approach to Biostatistics and Health Sciences

25 Bayesian Analysis and Prediction of Patients' Demands for Visits in Home Care 129
Raffaele Argiento, Alessandra Guglielmi, Ettore Lanzarone, and Inad Nawajah

26 Exploiting Adaptive Bayesian Regression Shrinkage to Identify Exome Sequence Variants Associated with Gene Expression 135
E.M. Boggis, M. Milo, and K. Walters

27 Randomized Phase II Trials: A Bayesian Two-Stage Design 139
Matteo Cellamare, Valeria Sambucini, and Federica Siena

28 Bayesian Matrix Factorization for Outlier Detection: An Application in Population Genetics 143
Nicolas Duforet-Frebourg and Michael G.B. Blum

29 Noise Model Selection for Multichannel Diffusion-Weighted MRI 149
Edward Knock, Theodore Kypraios, Paul Morgan, and Stamatios Sotiropoulos

30 Analysis of Hospitalizations of Patients Affected by Chronic Heart Disease ... 155
Alice Parodi, Francesca Ieva, Alessandra Guglielmi, and Raffaele Argiento

31 A Semiparametric Bayesian Multivariate Model for Survival Probabilities After Acute Myocardial Infarction ... 161
Elena Prandoni, Alessandra Guglielmi, Francesca Ieva, and Anna Maria Paganoni

32 Particle Learning Approach to Bayesian Model Selection: An Application from Neurology ... 165
Simon Taylor, Gareth Ridall, Chris Sherlock, and Paul Fearnhead

Part V Bayesian Models for Stochastic and Economic Processes

33 Analysis of Italian Financial Market via Bayesian Dynamic Covariance Models ... 171
Daniele Durante

34 Bayesian Model Selection of Regular Vine Copulas ... 177
Lutz F. Gruber and Claudia Czado

35 Analysis of Exchange Rates via Multivariate Bayesian Factor Stochastic Volatility Models ... 181
Gregor Kastner, Sylvia Frühwirth-Schnatter, and Hedibert F. Lopes

36 On Some Stationary Models: Construction and Estimation ... 187
Consuelo R. Nava, Ramsés H. Mena, and Igor Prünster

37 Claim Sizes in the Compound Poisson Process from a Bayesian Viewpoint ... 193
Gamze Özel

38 Land Rental Market and Agricultural Production Efficiency: A Bayesian Perspective ... 197
Haoran Yang

Part VI Suggestions for Young Researchers

39 The Point Is...to Publish? ... 203
Fulvia Mecatti

Part I
Theoretical Bayes

Chapter 1
A Nonparametric Model for Stationary Time Series

Isadora Antoniano-Villalobos and Stephen G. Walker

Abstract We present a family of autoregressive models with nonparametric stationary and transition densities, which achieve substantial modelling flexibility while retaining desirable statistical properties for inference. Posterior simulation involves an intractable normalizing constant; we therefore present a latent extension of the model which enables exact inference through a trans-dimensional MCMC method. We argue the capacity of this family of models to capture time homogeneous transition mechanisms, making them a powerful tool for predictive inference even when the process generating the data does not have a stationary density. Numerical illustrations are presented.

1.1 Introduction

The mixture of Dirichlet process (MDP) model, introduced by Lo [5], is a very popular model, which has benefitted from the advances in simulation techniques, so that the model is now able to cover more complex data structures, such as regression models and time series models [3].

In the context of time series, there is a need for flexible models which can accommodate complex dynamics, observed in real-life data. While stationarity is a desirable property, which facilitates estimation of relevant quantities, it is difficult to construct stationary models for which both the transition mechanism

---

I. Antoniano-Villalobos (✉)
Department of Decision Sciences, Bocconi University, Milan, Italy
e-mail: isadora.antoniano@unibocconi.it

S.G. Walker
Department of Mathematics, University of Texas, Austin, USA
e-mail: s.g.walker@math.utexas.edu

and the invariant density are sufficiently flexible. Many attempts have been made, often resulting in a compromise between flexibility and statistical properties (see, e.g., [2, 6–9]).

We propose a model with nonparametric transition and stationary densities, which enjoys the advantages associated with stationarity, while retaining the necessary flexibility for both the transition and stationary densities. We demonstrate how posterior inference via MCMC can be carried out, focusing on the estimation of the transition density, both for stationary and non-stationary data-generating processes. For ease of exposition, we only consider first-order time series data and models, but the construction we propose can be adapted for higher-order Markov dependence structures.

1.2 The Model

We construct a nonparametric version of the usual autoregressive model by defining a nonparametric, i.e., infinite, mixture of parametric bivariate densities $K_\theta(y, x)$, for with both marginals, $K_\theta(y)$ and $K_\theta(x)$ are the same. We then define the transition density as the conditional density for y given x, therefore preserving the stationarity.

The transition mechanism can be expressed as a nonparametric mixture of transition densities with dependent weights,

$$f_P(y|x) = \sum_{j=1}^{\infty} w_j(x) K_{\theta_j}(y|x), \tag{1}$$

where

$$w_j(x) = \frac{w_j K_{\theta_j}(x)}{\sum_{j'=1}^{\infty} w_{j'} K_{\theta_{j'}}(x)}. \tag{2}$$

Clearly, the expression in the denominator, namely

$$f_P(x) = \sum_{j=1}^{\infty} w_j K_{\theta_j}(x), \tag{3}$$

constitutes the invariant density for such transition.

Therefore, both the transition and the stationary densities for the model are defined as nonparametric mixtures.

To our knowledge, the only other fully nonparametric Bayesian model for stationary Markov processes developed so far is due to Martínez-Ovando and Walker [6]. No applications to real data are currently available in the literature, probably due to the complex nature of their model construction.

The model we propose has a simple structure. However, it has been, until now, considered intractable due to the infinite mixture appearing in the denominator of the dependent weight expression. We therefore propose a latent variable extension

1 A Nonparametric Model for Stationary Time Series

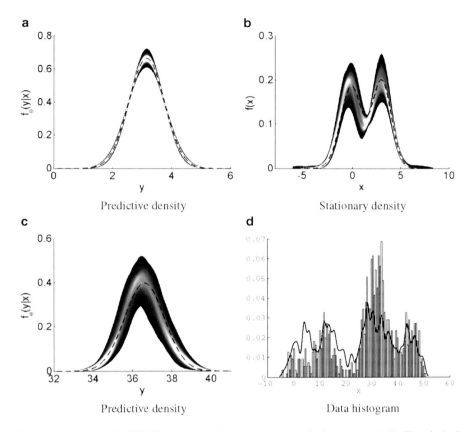

Fig. 1.1 Inference via MCMC estimation for two data sets of size $n = 1,000$. The *dashed lines* correspond to the true densities, which are accurately captured by point-wise estimates, represented by the *gray-scale map*, with highest posterior probability in *white*. (**a**) Mixture model data: transition density. (**b**) Mixture model data: stationary density. (**c**) Brownian motion data: transition density. (**d**) Brownian motion data: the marginal point estimate (*solid line*) provided by the model captures the variability of the data histogram, enabling transition density estimation even in the absence of a true stationary density

which enables posterior inference for the model via MCMC, involving slice sampling [4] and a trans-dimensional MCMC method [1]. Future work should include applications to real data.

1.2.1 Illustrations

We present some examples, all of them involving simulated data: form the mixture model itself, a stationary diffusion process, standard Brownian motion and a non-stationary diffusion. They illustrate how our model can be used for transition and invariant density estimation simultaneously, when the stationary density exists, yet remaining suitable for transition density estimation, even when the data is not generated by a stationary process (Fig. 1.1).

References

1. Godsill SJ (2001) On the relationship between Markov chain Monte Carlo methods for model uncertainty. J Comput Graph Stat 10(2):230–248
2. Griffin JE, Steel MFJ (2011) Stick-breaking autoregressive processes. J Econ 162(2):383–396
3. Hjort NL, Holmes C, Müller P, Walker SG (eds) (2010) Bayesian nonparametrics. Cambridge University Press, Cambridge
4. Kalli M, Griffin JE, Walker SG (2011) Slice sampling mixture models. Stat Comput 21:93–105
5. Lo AJ (1984) On a class of Bayesian nonparametric estimates: I. Density estimates. Ann Stat 12(1):351–357
6. Martínez-Ovando JC, Walker SG (2011) Time-series modelling, stationarity and Bayesian nonparametric methods. Technical report, Banco de México
7. Mena RH, Walker SG (2005) Stationary autoregressive models via a Bayesian nonparametric approach. J Time Ser Anal 26(6):789–805
8. Müller P, West M, MacEachern S (1997) Bayesian models for non-linear auto-regressions. J Time Ser Anal 18:593–614
9. Tang Y, Ghosal S (2007) A consistent nonparametric Bayesian procedure for estimating autoregressive conditional densities. Comput Stat Data Anal 51:4424–4437

Chapter 2
Estimation of Optimally Combined-Biomarker Accuracy in the Absence of a Gold Standard Reference Test

Leandro Garcia Barrado, Elisabeth Coart, and Tomasz Burzykowski

Abstract The reference diagnostic test used to establish the discriminative properties of a combination of biomarkers could be imperfect. This may lead to a biased estimate of the accuracy of the combination. A Bayesian latent-class mixture model is proposed to estimate the area under the ROC curve (AUC) of a combination of biomarkers. The model allows selecting the combination that maximizes the AUC and takes possible errors in the reference test into account. A simulation study was performed based on 400 data sets. Sample sizes from 100 to 600 observations were considered. Informative as well as non-informative prior information for the diagnostic accuracy of the reference test was considered. In addition, a controlled prior specification is proposed. The obtained average estimates for all parameters were close to the true values; some differences in efficiency were observed. Results indicate an adequate performance of the model-based estimates.

2.1 Introduction

Biomarkers can be used for developing a diagnostic test for a disease. Often, to increase the diagnostic accuracy of the test, a combination of several biomarkers is considered [6]. To assess the diagnostic performance of a biomarker-based test,

---

L.G. Barrado (✉)
Interuniversity Institute for Biostatistics and statistical Bioinformatics (I-BioStat),
Hasselt University, Agoralaan building D, 3590 Diepenbeek, Belgium
e-mail: leandro.garciabarrado@uhasselt.be

E. Coart
International Drug Development Institute (IDDI), Avenue Provinciale 30,
1340 Louvain-la-Neuve, Belgium
e-mail: elisabeth.coart@iddi.com

T. Burzykowski
I-BioStat, IDDI, Louvain-la-Neuve, Belgium
e-mail: tomasz.burzykowski@uhasselt.be

a so-called reference test, establishing disease status of an individual, is needed. Depending on the disease of interest, the reference test may be imperfect, i.e., it may misclassify the control and diseased individuals. In such a case, the estimate of the diagnostic accuracy of a biomarker could be biased [4]. Therefore, when developing a biomarker-based diagnostic test, the possibility of an imperfect reference test has to be taken into account.

2.2 Methods

We use the area under the ROC curve (AUC) as a measure of the diagnostic accuracy and a model to derive the linear combination of biomarkers maximizing the AUC [3]. In particular, a Bayesian latent-class mixture model is fitted to obtain estimates of the distributional parameters of the multivariate distributions for the biomarkers that form the mixture components for the diseased and control populations. By estimating the latent true disease status and component parameters through a mixture model with both reference test and biomarker values contributing to the likelihood, the misclassification probabilities of the reference test are taken into account [2].

A simulation study was performed to investigate the performance of the model under several settings for 400 simulated data sets. Sample sizes of 100, 400, and 600 observations were considered, split equally between the diseased and control groups. The prior distributions for the sensitivity and specificity of the reference test were varied from non-informative to informative.

Priors for the remaining parameters were set as standard non-informative priors [1]. This "naive" approach does not enable control of the prior distributions for the variances, correlations, and the AUC. Therefore, an alternative prior specification is proposed. Re-parametrization of the variance-covariance matrices allows a more direct specification of the prior information for variances and correlations [5]. By putting a prior distribution on the difference of the mixture component means, scaled by the sum of the variance-covariance matrices, prior information for the AUC can be precisely specified.

2.3 Results

Table 2.1 presents the results of the simulation study for the three considered sample sizes. The table contains the average of the median posterior AUC estimates over the 400 data sets for each of the different simulation settings. The rows marked Naive correspond to the "naive" approach, which leads to the AUC prior as in Fig. 2.1a. The rows marked Controlled correspond to the re-parametrized approach,

2 Estimation of Biomarker Accuracy in the Absence of a Gold Standard

Table 2.1 Mean (standard error) of the median posterior AUC of all 400 fits for all considered settings

Prior formulation	Se/Sp prior	True AUC	Sample size N = 100	N = 400	N = 600
Naive	Non-Inf	0.8786	0.9241 (0.0279)	0.8890 (0.0279)	0.8836 (0.0262)
Naive	Inf	0.8786	0.9068 (0.0344)	0.8827 (0.0286)	0.8785 (0.0263)
Controlled	Non-Inf	0.8786	0.8907 (0.0347)	0.8803 (0.0290)	0.8773 (0.0271)
Controlled	Inf	0.8786	0.8728 (0.0388)	0.8741 (0.0292)	0.8722 (0.0269)

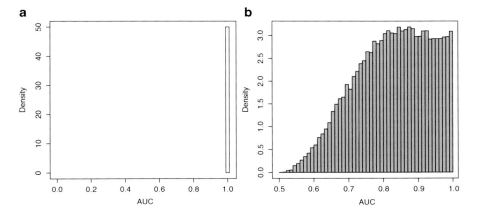

Fig. 2.1 Simulated implied priors for AUC based on mixture component priors. (**a**) Implied prior for the naive prior specification. (**b**) Implied prior for the proposed controlled prior specification

with the AUC prior shown in Fig. 2.1b. Within each parametrization approach, the rows indicated by Non-Inf and Inf represent the results for the non-informative and informative prior for the accuracy of the reference test, respectively.

Considering the naive approach, the results point to overestimation and decreasing efficiency of posterior estimates with decreasing sample size. Increasing the amount of prior information for the accuracy of the reference test resolves, or at least reduces, the bias observed for small data sets. Counterintuitively, the increase of the prior information leads to a decrease in efficiency of the AUC estimates.

It appears that the consequence of assuming non-informative priors for the parameters of the biomarker-related distributions is that the prior for the AUC essentially becomes a point mass distribution at one (see panel A of Fig. 2.1). This explains the overestimation of the AUC as due to the highly informative AUC prior distribution. Changing the parametrization of the model allows specifying the prior distribution for the AUC as in Fig. 2.1b. As a consequence, introducing this less informative prior reduces the bias. Increasing sample size or the amount of information in the prior for the accuracy of the reference test does not alter the results substantially, as shown in Table 2.1.

2.4 Conclusions

The results indicate that the model does provide unbiased estimates of the accuracy of the optimal combination of diagnostic biomarkers, but care has to be taken in the specification of the prior information, especially for the AUC. Under the "naive" approach, an informative prior for the accuracy of the imperfect reference test may overcome the informative prior for the AUC in small data sets.

References

1. O'Malley AJ, Zou KH (2006) Bayesian multivariate hierarchical transformation models for ROC analysis. Stat Med 25:459–479
2. Scott AN, Joseph L, Bélisle MA, Behr KS (2007) Bayesian modelling of tuberculosis clustering from DNA fingerprint data. Stat Med 27:140–156
3. Su JQ, Liu JS (1993) Linear combinations of multiple diagnostic markers. J Am Stat Assoc 88:1350–1355
4. Valenstein PN (1990) Evaluating diagnostic tests with imperfect standards. Am J Clin Pathology 93:252–258
5. Wei Y, Higgins PT (2013) Bayesian multivariate meta-analysis with multiple outcomes. Stat Med. doi:10.1002/sim.5745
6. Zhou XH, Obuchowski NA, McClish DK (2002) Statistical Methods in Diagnostic Medicine. Wiley, New York

Chapter 3
On Bayesian Transformation Selection: Problem Formulation and Preliminary Results

E. Charitidou, D. Fouskakis, and I. Ntzoufras

Abstract The problem of transformation selection is thoroughly treated from a Bayesian perspective. Several families of transformations are considered with a view to achieving normality: the *Box-Cox*, the *Modulus*, the *Yeo and Johnson* and the *Dual* transformation. Markov Chain Monte Carlo algorithms have been constructed in order to sample from the posterior distribution of the transformation parameter λ_T associated with each competing family T. We investigate different approaches to constructing compatible prior distributions for λ_T over alternative transformation families, using the power-prior and the unit-information prior approaches. In order to distinguish between different transformation families, posterior model probabilities have been calculated. Using simulated datasets, we show the usefulness of our approach.

3.1 Introduction

In the literature, the term *transformation selection* so far pertains to the choice of an optimal value for the transformation parameter within a given family. We introduce a two-step approach where a transformation family is selected at an initial level, while at a second level the value of the transformation parameter is specified given the family. Working within the Bayesian context requires careful choice of prior distributions. In our case, this becomes even more complex since the prior distribution for the transformation parameter λ_T under each family T needs to be compatible to account for the different interpretation of λ_T given T.

E. Charitidou (✉) • D. Fouskakis
Department of Mathematics, National Technical University of Athens, 15780 Athens, Greece
e-mail: echarit@central.ntua.gr; fouskakis@math.ntua.gr

I. Ntzoufras
Department of Mathematics, Athens University of Economics and Business, Athens 10434, Greece
e-mail: ntzoufras@aueb.gr

3.2 Bayesian Formulation

Four uniparametric families of transformations are considered and compared with each other: *Box-Cox* [1], *Modulus* [4], *Yeo and Johnson* [6] and *Dual* [5].

Each family is indexed by T and involves a transformation parameter λ_T. Let us denote by $\mathbf{y} = (y_1, \ldots, y_n)^T$ the observed data and by $\mathbf{y}^{(\lambda_T)} = \left(y_1^{(\lambda_T)}, \ldots, y_n^{(\lambda_T)}\right)^T$ the transformed data for a given value of the parameter λ_T within a particular transformation family T. We aim for $\mathbf{y}^{(\lambda_T)}$ to be a sample from a Normal distribution $N(\mu_T, \sigma_T^2)$ with unknown parameter vector (μ_T, σ_T^2) under some appropriate value of λ_T.

Regarding the prior probability of each of the six transformation families (including the identical and the log transformation), no special prior weight is assigned to any family, i.e., $\pi(T) = \frac{1}{|\mathscr{T}|} = \frac{1}{6}$. As to the prior on the transformation parameters, it has a hierarchical form $\pi(\boldsymbol{\theta}_T|T) = \pi(\mu_T, \sigma_T^2|\lambda_T, T)\pi(\lambda_T|T)$.

The main parameter of interest within a family is λ_T while (μ_T, σ_T^2) are regarded as nuisance parameters; therefore, we employ an independent Jeffreys prior (reference prior) for those.

On the grounds of the different interpretation of λ_T among families, the concept of the power-prior [3] is adopted in order to construct compatible prior distributions. The power-prior for λ_T is formed as the posterior distribution of a set of imaginary data $\mathbf{y}^*$ under a reference baseline prior $\pi_0(\lambda_T|T) \propto 1$:

$$\pi(\lambda_T|\mathbf{y}^*, T) = \frac{f(\mathbf{y}^*|\lambda_T, T)^{1/n^*}}{\int f(\mathbf{y}^*|\lambda_T, T)^{1/n^*} \, d\lambda_T}. \tag{1}$$

In addition, a unit-information Normal prior setting is used, based on the same imaginary data $\mathbf{y}^*$, which theoretically approximates the former power-prior setting. The variance of the latter prior is determined through the observed Fisher information given $\mathbf{y}^*$. Approximations of the intractable integrals included in the process are achieved through Chib's estimator [2] incorporating the output of a random-walk Metropolis-Hastings algorithm simulating from the marginal posterior distribution of λ_T.

3.3 Results

In order to illustrate our approach, we have simulated data from a variety of distributions. The Student distribution t_2, having non-centrality parameter equal to -1, is an example of particular interest since symmetry is accompanied by fat tails. The latter characteristic usually induces failure of transformation to normality under most families. Looking at the figures in Table 3.1, we observe that the supremacy of the Modulus family for this distribution is unquestionable for both medium and large sample sizes ($n = 100$ and $n = 1000$) under both power-prior and unit-information Normal prior.

3 On Bayesian Transformation Selection...

Table 3.1 Posterior model probabilities and log-marginal likelihood values for each transformation family T along with Monte Carlo estimates for the posterior median (sd) of λ_T, all estimated using Chib's approximation method for the Student simulated dataset

		Prior[a]	Modulus	Box-Cox	Dual	Id	YJ	Log
$n = 100$	$P(T\|y)$	Prior A	0.99	0.01	< 0.01	< 0.01	< 0.01	< 0.01
		Prior B	0.99	0.01	< 0.01	< 0.01	< 0.01	< 0.01
	$\log f(\mathbf{y}\|T)$	Prior A	−250.59	−255.89	−256.39	−258.28	−259.36	−292.14
		Prior B	−251.92	−257.01	−259.77	−258.28	−260.48	−292.14
	λ_T	Prior A	0.36 (0.15)	1.60 (0.22)	1.62 (0.22)	–	1.08 (0.08)	–
		Prior B	0.34 (0.15)	1.62 (0.23)	1.63 (0.21)	–	1.09 (0.08)	–
		Prior	Modulus	Box-Cox	Dual	YJ	Id	Log
$n = 1000$	$P(T\|y)$	Prior A	> 0.99	< 0.01	< 0.01	< 0.01	< 0.01	< 0.01
		Prior B	> 0.99	< 0.01	< 0.01	< 0.01	< 0.01	< 0.01
	$\log f(\mathbf{y}\|T)$	Prior A	−3,733.00	−3,960.67	−3,961.52	−4,027.56	−4,125.02	−4,601.98
		Prior B	−3,732.75	−3,960.92	−3,963.83	−4,027.63	−4,125.02	−4,601.98
	λ_T	Prior A	−0.01 (0.04)	2.93 (0.12)	2.93 (0.12)	1.23 (0.01)	–	–
		Prior B	−0.01 (0.04)	2.94 (0.12)	2.94 (0.12)	1.23 (0.01)	–	–

[a] Prior A: Unit-information Normal prior; Prior B: Power-prior

3.4 Conclusions

The compatibility issues in transformation selection have been addressed through the power-prior approach. By and large, there is more than adequate convergence of results under both prior settings. The fat-tailed Student distribution is optimally associated to the Modulus transformation. The latter result has been verified for other fat-tailed distributions such as the Laplace.

Acknowledgements This work has been partially funded by the Research Committee of the National Technical University of Athens (Π.E.B.E. 2010 Scheme).

References

1. Box GEP, Cox DR (1964) An analysis of transformations (with discussion). J R Stat Soc Ser B 26:211–252
2. Chib S, Jeliazkov I (2001) Marginal likelihood from the Metropolis-Hastings output. J Amer Statist Assoc 96:270–281
3. Ibrahim JG, Chen MH (2000) Power-prior distributions for regression models. Statist Sci 15: 46–60
4. John JA, Draper NR (1980) An alternative family of transformations. J R Stat Ser C 29:190–197
5. Yang Z (2006) A modified family of power transformations. Econ Lett 92:14–19
6. Yeo IK, Johnson RA (2000) A new family of power transformations to improve normality or symmetry. Biometrika 87:954–959

Chapter 4
A Simple Proof for the Multinomial Version of the Representation Theorem

Marcio A. Diniz and Adriano Polpo

Abstract In this work we present a demonstration for the multinomial version of de Finetti's Representation Theorem. We use characteristic functions, following his first demonstration for binary random quantities, but simplify the argument through forward operators.

4.1 Introduction

In 1928, Bruno de Finetti presented a contribution at the International Congress of Mathematicians: *Funzione caratteristica di un fenomeno aleatorio*, but it was published only in 1932, when the sixth and last volume of the annals of that congress was released. In 1930, a more detailed version was published, as a *memoir*, by the *Accademia dei Lincei*, with the same title.[1]

De Finetti used an analytic argument and characteristic functions, alongside with the exchangeability hypothesis, to prove his famous Representation Theorem. In this work we use his arguments with forward operators to present a new proof for the multinomial case.

---
[1] See [1,5,6].

M.A. Diniz (✉) • A. Polpo
Federal University of S. Carlos, Rod. Washington Luis, km 235, S. Carlos, Brazil
e-mail: marcio.diniz@ugent.be; polpo@ufscar.br

4.2 De Finetti's Method for Multinomial Trials

Let an infinite sequence of random quantities that assume any of k values or categories considered to be exchangeable. We want to study some subsequence of n of such quantities. In order to do this, we consider the vector $(S_1, \ldots, S_{k-1})$ that displays the number of quantities that assumed value $1, 2, \ldots, k-1$. The probability-generating function (p.g.f) of such vector is, for $z \in \mathbb{C}^k$:

$$\Omega_n(z_1, \ldots, z_{k-1}) = \sum_\Delta \omega^{(n)}_{h_1,\ldots,h_{k-1}} z_1^{h_1} \cdots z_{k-1}^{h_{k-1}} \tag{1}$$

in which $\Delta = \{(x_1, \ldots, x_{k-1}) \in \mathbb{Z}_+^{k-1} : x_1 + \cdots x_{k-1} \leq n\}$, $n \in \mathbb{N}$ and $\omega^{(n)}_{h_1,\ldots,h_{k-1}}$ is the probability that we observe, in n trials, h_1 of category $1, \ldots, h_{k-1}$ of category $k-1$, regardless of the order. We also denote by $\omega_0^{(0)} = 1$, $t = \sum_{i=1}^{k-1} h_i$ and

$$\omega^{(t)}_{h_1,\ldots,h_{k-1}} = \omega_{h_1,\ldots,h_{k-1}}.$$

Exchangeability makes it possible to write

$$\frac{\omega^{(n)}_{h_1,\ldots,h_{k-1}}}{\binom{n}{h_1,\ldots,h_{k-1}}} = \delta^{n-t} \omega_{h_1,\ldots,h_{k-1}} \geq 0 \tag{2}$$

in which the δ operator is defined after [3] by

$$\delta \omega_{h_1,\ldots,h_{k-1}} = \omega_{h_1,\ldots,h_{k-1}} - \omega_{h_1+1,\ldots,h_{k-1}} - \omega_{h_1,h_2+1,\ldots,h_{k-1}} - \cdots - \omega_{h_1,\ldots,h_{k-1}+1}.$$

Now we define the forward operators. Let us denote

$$F_j^k \omega_{h_1,\ldots,h_{k-1}} = \omega_{h_1,\ldots,h_j+k,\ldots,h_{k-1}}$$

then it follows that

$$F_i^r F_j^k \omega_{h_1,\ldots,h_{k-1}} = F_j^k F_i^r \omega_{h_1,\ldots,h_{k-1}} = \omega_{h_1,\ldots,h_i+r,\ldots,h_j+k,\ldots,h_{k-1}}$$

and that

$$\delta^{n-t} \omega_{h_1,\ldots,h_{k-1}} = (1 - F_1 - \ldots - F_{k-1})^{n-t} \omega_{h_1,\ldots,h_{k-1}}.$$

Using (2), implied by the exchangeability hypothesis and forward operators, the p.g.f (1) may be written as

4 A Simple Proof for the Multinomial Version of the Representation Theorem

$$\Omega_n(z_1,\ldots,z_{k-1}) = \sum_\Delta \omega_{h_1,\ldots,h_{k-1}}^{(n)} z_1^{h_1} \cdots z_{k-1}^{h_{k-1}}$$

$$= \sum_\Delta \binom{n}{h_1,\ldots,h_{k-1}} z_1^{h_1} \cdots z_{k-1}^{h_{k-1}} (1 - F_1 \cdots - F_{k-1})^{n-h_1\cdots-h_{k-1}} \omega_{h_1,\ldots,h_{k-1}}$$

$$= \sum_\Delta \binom{n}{h_1,\ldots,h_{k-1}} (z_1 F_1)^{h_1} \cdots (z_{k-1} F_{k-1})^{h_{k-1}} (1 - F_1 \cdots - F_{k-1})^{n-t} \omega_0$$

$$= [1 + F_1(z_1 - 1) + \cdots + F_{k-1}(z_{k-1} - 1)]^n \omega_0. \tag{3}$$

Following de Finetti's approach, we define the characteristic function of $\overline{S} = (S_1/n, \ldots, S_{k-1}/n)$ which, using (3), may be written as

$$\Psi_{\overline{S}}(t_1,\ldots,t_{k-1}) = \Omega_n(e^{it_1/n},\ldots,e^{it_{k-1}/n})$$

$$= [1 + F_1(e^{it_1/n} - 1) + \cdots + F_{k-1}(e^{it_{k-1}/n} - 1)]^n \omega_0$$

and study its limit when $n \to \infty$. It is possible to show[2] that

$$\Psi_\Theta(t_1,\ldots,t_{k-1}) = \exp[i(F_1 t_1 + F_2 t_2 + \cdots + F_{k-1} t_{k-1})] \omega_0 \tag{4}$$

and, by Levy's continuity theorem, it is known that (4) is the characteristic function of only one random vector, Θ, that assumes value in the $k - 1$ simplex and whose distribution function[3] is denoted Φ. Given the properties relating moments and characteristic functions, we can rewrite (1) once more, through the multinomial theorem

$$\Omega_n(z_1,\ldots,z_{k-1}) = \int_{\mathbb{S}^{k-1}} [1 + \theta_1(z_1 - 1) + \cdots + \theta_{k-1}(z_{k-1} - 1)]^n d\Phi(\theta)$$

$$= \sum_\Delta \binom{n}{h_1,\ldots,h_{k-1}} z_1^{h_1} \cdots z_{k-1}^{h_{k-1}} \int_{\mathbb{S}^{k-1}} \theta_1^{h_1} \theta_2^{h_2}$$

$$\ldots (1 - \theta_1 - \cdots - \theta_{k-1})^{n-t} d\Phi(\theta),$$

where $\theta \in \mathbb{S}^{k-1}$, the $(k - 1)$-simplex, and from it follows that

$$\omega_{h_1,\ldots,h_{k-1}}^{(n)} = \binom{n}{h_1,\ldots,h_{k-1}} \int_{\mathbb{S}^{k-1}} \theta_1^{h_1} \theta_2^{h_2} \cdots (1 - \theta_1 - \cdots - \theta_{k-1})^{n-t} d\Phi(\theta)$$

that is de Finetti's Representation Theorem for multinomial sequences of exchangeable random quantities.

---

[2] The limits with forward operators are well defined because the set of polynomial operators induces an algebra that is isomorphic to the algebra of polynomials in real or complex variables. See [4].
[3] It is possible to find the distribution function inverting the characteristic function.

De Finetti [7] does not provide a proof for the multinomial case but only asymptotical arguments that, starting from the finite binomial case, it is possible to derive the infinite multinomial case. For the binomial case, Bernardo and Smith [2] provide a proof based on [8], but for the multinomial case the proof is reported as "a straightforward, albeit algebraic cumbersome, generalization of the proof of (Representation theorem for binary random quantities)". The proof given by [8] can be considered as, essentially, de Finetti's proof with a limit argument not involving characteristic functions.

The results presented here provided a simple and clean demonstration of de Finetti's Representation Theorem for infinite sequences of multinomial random quantities.

Acknowledgements Marcio Diniz was supported by FAPESP (Sao Paulo Research Foundation), under the project 2012/14764-0, and wishes to thank SYSTeMs Research Group at Ghent University for its hospitality and support.

References

1. Bassetti F, Regazzini E (2008) The unsung de Finetti's first paper about exchangeability. Rendiconti di Matematica, Serie VII 28:1–17
2. Bernardo JM, Smith AF (1994) Bayesian theory. Wiley, New York
3. Dale AI (1987) Two-dimensional moment problems. Math Scientist 12:21–29
4. Dhrymes PJ (2000) Mathematics for econometrics, 3rd edn, Springer, New York
5. de Finetti B (1930) Funzione caratteristica di un fenomeno aleatorio. Memorie della Academia dei Lincei IV(5):86–133
6. de Finetti B (1932) Funzione caratteristica di un fenomeno aleatorio. Atti del Congresso Internazionale dei Matematici, Bologna, 3–10 Settembre 1928, pp 179–190
7. de Finetti B (1972) Probability, induction and statistics: the art of guessing. Wiley, New York
8. Heath DL, Sudderth W (1976) De Finetti's Theorem for exchangeable random variables. Am Statist 30:333–345

Chapter 5
A Sequential Monte Carlo Framework for Adaptive Bayesian Model Discrimination Designs Using Mutual Information

Christopher C. Drovandi, James M. McGree, and Anthony N. Pettitt

Abstract In this paper we present a unified sequential Monte Carlo (SMC) framework for performing sequential experimental design for discriminating between a set of models. The model discrimination utility that we advocate is fully Bayesian and based upon the mutual information. SMC provides a convenient way to estimate the mutual information. Our experience suggests that the approach works well on either a set of discrete or continuous models and outperforms other model discrimination approaches.

5.1 Introduction

The problem of model choice within a Bayesian framework has received an abundance of attention in the literature. Therefore, when a set of competing models is proposed a priori, it is important to determine the optimal selection of the controllable aspects (when available) of the experiment for discriminating between the models. A sequential experimental design allows experiments to be performed in batches, so that adaptive decisions can be made for each new batch.

In this paper we adopt a unified sequential Monte Carlo (SMC) framework for performing model discrimination in sequential experiments. We consider as a utility the mutual information between the model indicator and the next observation(s) [1]. SMC allows for convenient estimation of posterior model probabilities [3] as well as the mutual information utility, both of which are generally difficult to calculate.

C.C. Drovandi (✉) • J.M. McGree • A.N. Pettitt
Queensland University of Technology, GPO Box 2434, Brisbane 4001, Australia
e-mail: c.drovandi@qut.edu.au; james.mcgree@qut.edu.au; a.pettitt@qut.edu.au

In SMC, new data can be accommodated via a simple re-weighting step. Thus, the simulation properties of various utilities can be discovered in a timely manner with SMC compared with approaches that use Markov chain Monte Carlo to recompute posterior distributions (see [5]).

From our experience we have found the approach to be successful on several diverse applications, including models for both discrete (see [4]) and continuous (see [7]) data. The purpose of this paper is to collate [4, 7] into a single source describing the SMC mutual information for model discrimination calculation for applications involving a set of discrete or continuous models. Section 5.2 develops the notation, Sect. 5.3 details SMC under model uncertainty and Sect. 5.4 describes the mutual information calculation. Section 5.5 describes the examples our approach has been tested on while Sect. 5.6 concludes the paper.

5.2 Notation

We use the following notation. We consider a finite number of K models, described by the random variable $M \in \{1, \ldots, K\}$. We assume one of the K models is responsible for data generation. Each model m contains a parameter, $\theta_\mathbf{m}$, with a likelihood function, $f(\mathbf{y_t}|m, \theta_\mathbf{m}, \mathbf{d_t})$, where $\mathbf{y_t}$ represents the data collected up to current time t based on the selected design points, $\mathbf{d_t}$. We place a prior distribution over $\theta_\mathbf{m}$ for each model, denoted by $\pi(\theta_\mathbf{m}|m)$. $\pi(m)$ and $\pi(m|\mathbf{y_t}, \mathbf{d_t})$ are the prior and posterior probability of model m, respectively.

5.3 Sequential Monte Carlo Incorporating Model Uncertainty

SMC consists of a series of re-weighting, re-sampling and mutation steps. For a single model, we use the algorithm of [2]. For sequential designs involving model uncertainty, we run SMC algorithms in parallel for each model and combine them after introducing each observation to compute posterior model probabilities and the mutual information utility. We denote the particle set at target t for the mth model obtained by SMC as $\{W_{m,t}^i, \theta_{\mathbf{m,t}}^\mathbf{i}\}_{i=1}^N$, where N is the number of particles. It is well known that SMC provides a simple way to estimate the evidence for a particular model based on importance weights, which can be converted to estimates of the posterior model probabilities. The reader is referred to [4] for more details on the algorithm.

5.4 Mutual Information for Model Discrimination

For model discrimination, we advocate the use of the mutual information utility between the model indicator and the next observation, first proposed in [1]. This utility provides us with the expected gain in information about the model indicator introduced by the next observation. In general it is difficult to calculate; however, SMC allows efficient calculation. One can show that the utility for the design d to apply for the next observation z is given by

$$U(d|\mathbf{y_t}, \mathbf{d_t}) = \sum_{m=1}^{K} \pi(m|\mathbf{y_t}, \mathbf{d_t}) \int_{z \in \mathscr{S}} f(z|m, \mathbf{y_t}, \mathbf{d_t}, d) \log \pi(m|\mathbf{y_t}, \mathbf{d_t}, z, d) dz, \quad (1)$$

where $\mathscr{S}$ is the sample space of the response z. Below, we denote SMC estimates of predictive distributions and posterior model probabilities with a hat. If z is discrete, a summation replaces the integral

$$\hat{U}(d|\mathbf{y_t}, \mathbf{d_t}) = \sum_{m=1}^{K} \hat{\pi}(m|\mathbf{y_t}, \mathbf{d_t}) \sum_{z \in \mathscr{S}} \hat{f}(z|m, \mathbf{y_t}, \mathbf{d_t}, d) \log \hat{\pi}(m|\mathbf{y_t}, \mathbf{d_t}, z, d), \quad (2)$$

[4]. When z is continuous, the integral can be approximated using the SMC particle population for each model

$$\hat{U}(d|\mathbf{y_t}, \mathbf{d_t}) = \sum_{m=1}^{K} \hat{\pi}(m|\mathbf{y_t}, \mathbf{d_t}) \sum_{i=1}^{N} W_{m,t}^i \log \hat{\pi}(m|\mathbf{y_t}, \mathbf{d_t}, z_{m,t}^i, d), \quad (3)$$

[7] where $z_{m,t}^i \sim f(z|m, \theta_{m,t}^i, d)$ if the observations are independent.

5.5 Examples

The SMC algorithm for designing in the presence of model uncertainty together with the use of the mutual information utility function has been tested on a variety of discrete and continuous model examples spanning several application areas. The SMC algorithm facilitated faster assessment of different utility functions for model discrimination purposes. Drovandi et al. [4] considered binary and count data examples. The applications included memory retention models, dose-response relationships in the context of clinical trials and models for neuronal degeneration. In all cases the mutual information utility led to a more rapid identification of the correct model compared to a random design. McGree et al. [7] applied the algorithm to continuous model examples. The methodology was illustrated on competing models for an asthma dose-finding study, a chemical engineering application and a pharmacokinetics example. The mutual information utility was compared to a random design and the total separation criterion (see, e.g., [6]), which is another model discrimination utility. We found that the mutual information utility led to designs that were more robust for detecting the correct model across applications.

5.6 Conclusion

Here we have brought together the findings of [4, 7] into a single source for performing adaptive Bayesian model discrimination under discrete or continuous model uncertainty. The methodology relies on SMC, which has already proven to be useful in sequential designs [5] and furthermore provides a convenient estimate of the mutual information utility we advocate for model discrimination. The combination of the SMC algorithm and mutual information utility has been successfully tested on a wide range of applications.

References

1. Box GEP, Hill WJ (1967) Discrimination among mechanistic models. Technometrics 9:57–71
2. Chopin N (2002) A sequential particle filter method for static models. Biometrika 89:539–551
3. Del Moral P, Doucet A, Jasra A (2006) Sequential Monte Carlo samplers. J Roy Stat Soc Ser B Stat Methodol 68:411–436
4. Drovandi CC, McGree JM, Pettitt AN (2012) A sequential Monte Carlo algorithm to incorporate model uncertainty in Bayesian sequential design. J Comput Graph Stat. doi:10.1080/10618600.2012.730083
5. Drovandi CC, McGree JM, Pettitt AN (2013) Sequential Monte Carlo for Bayesian sequentially designed experiments for discrete data. Comput Stat Data Anal 57:320–335
6. Masoumi S, Duever TA, Reilly PM (2013) Sequential Markov chain Monte Carlo (MCMC) model discrimination. Cand J Chem Eng 91:862–869
7. McGree JM, Drovandi CC, Pettitt AN (2013) A sequential Monte Carlo approach to the sequential design for discriminating between rival continuous data models. http://eprints.qut.edu.au/53813/.

Chapter 6
Joint Parameter Estimation and Biomass Tracking in a Stochastic Predator–Prey System

Laura Martín-Fernández, Gianni Gilioli, Ettore Lanzarone,
Joaquín Míguez, Sara Pasquali, Fabrizio Ruggeri, and Diego P. Ruiz

Abstract A Rao–Blackwellized particle filter for estimating the behavioral parameter of the functional response and tracking the biomass of each population in a stochastic predator–prey system is presented in this paper. We consider a predator–prey model with a Lotka–Volterra functional response and small sets of field data. A first validation of the approach has been carried out using synthetic data.

6.1 Introduction

Successful establishment of biological control strategies is difficult because the current abundance of pest population and properties of the predator functional response, i.e., the per capita rate of predation, should be known, but this information is not always available. Moreover, the decision on time and amount of predator released has to be taken into the dynamical framework of predator–prey interaction.

L. Martín-Fernández (✉) • D.P. Ruiz
Departamento de Física Aplicada, Universidad de Granada, Granada, Spain
e-mail: lauramartin@ugr.es; druiz@ugr.es

G. Gilioli
Dipartimento di Scienze Biomediche e Biotecnologie, Università di Brescia, Brescia, Italy
e-mail: gianni.gilioli@med.unibs.it

E. Lanzarone • S. Pasquali • F. Ruggeri
CNR-IMATI, Milano, Italy
e-mail: ettore.lanzarone@cnr.it; sara.pasquali@mi.imati.cnr.it; fabrizio.ruggeri@mi.imati.cnr.it

J. Míguez
Departamento de Teoría de la Señal y Comunicaciones, Universidad Carlos III de Madrid, Madrid, Spain
e-mail: jmiguez@ieee.org

In this paper, we propose a method for the joint estimation of the dynamical biomass of each population and the feeding rate during the time evolution of population interactions.

6.2 Method

6.2.1 State-Space Model

We consider the nonlinear state-space model, obtained by the Euler discretization of a stochastic Lotka–Volterra type of model based on [4],

$$x_{k+1} = x_k + \tau \left[r x_k (1 - x_k) - q_0 x_k y_k \right] - \sigma x_k y_t \Delta w_{k+1}^{(1)} + \varepsilon x_k \Delta w_{k+1}^{(2)},$$

$$y_{k+1} = y_k + \tau \left[c q_0 x_k y_k - u y_k \right] + c \sigma x_k y_k \Delta w_{k+1}^{(1)} + \eta y_k \Delta w_{k+1}^{(3)},$$

$$o_{k+1}^x \sim \Gamma(x_{k+1}, d_x^2),$$

$$o_{k+1}^y \sim \Gamma(y_{k+1}, d_y^2), \tag{1}$$

where x_{k+1} and y_{k+1} are the biomass of prey and predator, respectively, at time $k+1$ per habitat unit normalized with respect to the prey carrying capacity per habitat unit (plant), o_{k+1}^x and o_{k+1}^y are noisy biomass observations defined as Gamma variables with mean equal to x_{k+1} and y_{k+1} and variance equal to d_x^2 and d_y^2, respectively, τ is the time step used in the Euler approximation, and $k = 0, 1, \ldots, S$ denotes the discrete time instants. The parameters r, c and u are species-specific and have been estimated in [4]. The increments of the Wiener processes, $\Delta w_{k+1}^{(1)}$, $\Delta w_{k+1}^{(2)}$ and $\Delta w_{k+1}^{(3)}$ are independent Gaussian variables with zero mean and variance τ, and the parameters σ, ε, and η have been estimated in [4].

Assume that the parameter q_0 in the functional response $q_0 x_t y_t$ is unknown and the goal is the joint estimation of this behavioral parameter and the biomass variables x_{k+1} and y_{k+1}.

6.2.2 Rao–Blackwellized Particle Filter

We apply a practical particle filter (PF) to approximate the sequence of posterior probability distributions of the biomass of each population with unknown parameter q_0 given the observations. The proposed algorithm is an example of a Rao–Blackwellized particle filter (RBPF) [2, 3]. Conditional on the sequences $\mathbf{x}_{0:k}$ and $\mathbf{y}_{0:k}$, the estimation of q_0 is solved numerically using a simple Kalman filtering algorithm [1, 5]. The RBPF handles a set of M particles in the two-dimensional space of the prey and predator biomass and a bank of M Kalman filters running in parallel.

This particle filter method is adapted to small observation datasets, updating importance weights and resampling the particle set only when experimental observations become available.

6.3 Experimental Results

6.3.1 Dataset Simulation

We consider the acarine predator–prey system studied in [4], the prey mite *Tetranychus urticae*, and the predator mite *Phytoseiulus persimilis*. The population dynamics is described by Eq.(1) where all parameters are defined in [4] and the behavioral parameter q_0 is unknown.

In order to generate a synthetic dataset, we set $q_0 = 1.9$, a time period $\tau = 1$ day, and a final time $S = 69$ days. Then we use the model in Eq.(1) to generate sequences of normalized prey $x_{1:S}$ and predator $y_{1:S}$ population biomass values. From these complete sequences, we generate eight noisy observations with variance $d_x^2 = d_y^2 = 10^{-4}$.

6.3.2 Validation of the RBPF Algorithm

We apply the RBPF algorithm with $M = 10^5$ particles to jointly estimate the unknown parameter q_0 and track the prey and predator biomass given the available set of eight synthetic observations. All particles are initialized in the same way, $x_0 = 0.1$ and $y_0 = 0.01$ are set, and a Gaussian distribution[1] is assumed for the prior density of q_0 with zero mean and variance equal to one.

Figure 6.1 shows the online evaluation of the posterior mean of q_0 generated by the RBPF method. At the final time $S = 69$, the value of the posterior mean converges to 1.946 and the posterior variance is 0.025.

For the same simulation, Fig. 6.1 also displays the true (synthetic) sequences $x_{0:S}$ and $y_{0:S}$ together with the online biomass estimates. It can be seen that the estimates are accurate at the times where observations are processed, but there is a drift (the error increases) when data are not available, especially for $k < 40$. We also see that for $k \geq 40$ the estimates of q_0 are more accurate, and this also affects the accuracy of the biomass estimation.

---

[1]The proposed methods demand that the prior of q_0 be Gaussian for formal consistency. However, even with the mean of q_0 at $k = 0$ equal to zero, the inference algorithm performs well; hence, we have chosen to use this prior to illustrate the robustness of the method.

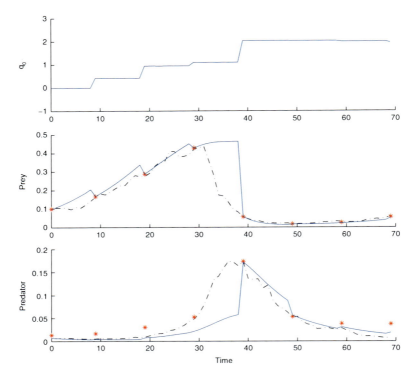

Fig. 6.1 Estimates of the unknown parameter q_0 over time and comparison of the true synthetic biomass sequences (*dash-dotted lines*) and the online biomass estimates (*continuous lines*) generated by the RBPF algorithm. The points for which observations are available are displayed with stars

6.4 Conclusions

Within the adaptive management framework in Integrated Pest Management (IPM), the predator–prey model we propose can undergo changes leading to improved predictive and explicative capabilities as more information becomes available. Differently from MCMC methods, the PF method does not present restriction on the dataset. In fact, it can be applied also during the period of data collection without waiting up to the end of at least one cycle of the population like in [4].

Compared to standard particle filters, the proposed method reduces both the dimension of the state space and the variance of the resulting estimates [2, 3].

The use of the Lotka–Volterra model implies an unsaturated capability of prey biomass intake for the predator. However, the intrinsic limitation in the model is outpaced by the advantages offered by the availability of prompt and progressively improved estimation of the predator functional response.

Finally, the experimental results confirm the goodness of the proposed method. As future work, an application to collected field data will be considered.

Acknowledgements This work has been supported by the "Consejería de Innovación, Ciencia y Empresa de la Junta de Andalucía" of Spain under project TIC-03269 and "Ministerio de Economía y Competitividad" of Spain under project COMPREHENSION TEC2012-38883-C02-01.

References

1. Anderson BDO, Moore JB (1979) Optimal filtering. Englewood Cliffs
2. Chen R, Liu JS (2000) Mixture Kalman filters. J Roy Stat Soc B 62:493–508
3. Doucet A, Godsill S, Andrieu C (2000) On sequential Monte Carlo sampling methods for Bayesian filtering. Stat Comput 10(3):197–208
4. Gilioli G, Pasquali S, Ruggeri F (2008) Bayesian inference for functional response in a stochastic predator-prey system. Bull Math Biol 70:358–381
5. Kalman RE (1960) A new approach to linear filtering and prediction problems. J Basic Eng 82:35–45

Chapter 7
Adaptive Bayes Test for Monotonicity

Jean-Bernard Salomond

Abstract We study the asymptotic behavior of a Bayesian nonparametric test of qualitative hypotheses. More precisely, we focus on the problem of testing monotonicity of a regression function. Even if some results are known in the frequentist framework, no Bayesian testing procedure has been proposed; at least none has been studied theoretically. This paper proposes a procedure that is straightforward to implement, which is a great advantage compared to those proposed in the literature.

7.1 Introduction

Shape-constrained models are of growing interest in the nonparametric field. Among them monotonicity constrains are very popular. There is a wide literature on the problem of estimating a function under monotonicity constrains. Groeneboom [7] and [14, 15] among others study the nonparametric maximum likelihood estimator of monotone densities; Brunner et al. [5, 10, 11, 18] study the properties of a Bayesian estimator. Barlow et al. [3] and [12] proposed a shape constrain estimator of monotonic regression functions. These methods are widely applied in practice. Bornkamp and Ickstadt [4] consider monotone function when modeling the response to a drug as a function of the dose and Hussain et al. [9] use a monotone representation for environmental data.

In this work, we propose a procedure to test for monotonicity constrains. We consider the Gaussian regression model

$$Y_i = f(i/n) + \epsilon_i, \ \epsilon_i \stackrel{iid}{\sim} \mathcal{N}\left(0, \sigma^2\right), \sigma^2 > 0, \ i = 1, \ldots, n, \qquad (1)$$

J.-B. Salomond (✉)
CREST and Université Paris Dauphine, 3 avenue Pierre Larousse 92245 Malakoff, France
e-mail: jean.bernard.salomond@ensae.fr

and, with $\mathscr{F}$ being the set of all monotone function, we test

$$H_0 : f \in \mathscr{F}, \text{ versus } H_1 : f \notin \mathscr{F}. \tag{2}$$

Thus both the null and the alternative are nonparametric hypotheses. The problem of testing for monotonicity has already been addressed in the frequentist literature and a variety of approaches have been considered. Baraud et al. [2] use projections of the regression function on the sets of piecewise constant function on a collection of partition of support of f. Their test rejects monotonicity if there is at least one partition such that the estimated projection is too far from the set of monotone functions. Another approach, considered in [6, 8] among others, is to test for negativity of the derivative of the regression function. However this requires some assumptions on the regularity of the regression function under the null hypothesis that could be avoided. In a recent paper Akakpo et al. [1] propose a procedure that detects local departure from monotonicity and study very precisely its asymptotic properties.

Here, we consider a Bayesian approach to this problem, which to the author's knowledge has not been studied. We only consider the case where $\mathscr{F}$ is the set of monotone nonincreasing functions, but a similar approach could be used when considering the set of monotone increasing or simply monotone functions. The most common approach to testing in a Bayesian setting is the Bayes factor. Here, however, we see that this method has drawbacks and seems to have poor performances.

Since monotone nonincreasing densities are well approximated by piecewise constants (see [7] or [18]), it is natural to build a prior on such functions. The standard approach to test for monotonicity of f in a Bayesian setting would be to consider the Bayes factor

$$B_{0,1} = \frac{\pi(f \in \mathscr{F}|Y^n)}{\pi(f \notin \mathscr{F}|Y^n)} \frac{1 - \pi(\mathscr{F})}{\pi(\mathscr{F})},$$

where π has the form for all $k \geq 2$ and all f written as

$$f = \sum_{i=1}^{k} \mathbb{I}_{[(i-1)/k, i/k)} \omega_i, \ d\pi(f) = \pi(k)\pi(\omega_1, \ldots, \omega_k|k).$$

However, this approach yields poor results in practice. The use of noninformative priors when considering the Bayes factor is not recommended as pointed out in [19]. We thus use a different approach and test

$$H_0^a : \tilde{d}(f, \mathscr{F}) \leq \tau_n \text{ versus } H_1^a : \tilde{d}(f, \mathscr{F}) > \tau_n, \tag{3}$$

where $\tilde{d}(f, \mathscr{F})$ is a distance between f and the set of monotone nonincreasing function $\mathscr{F}$ and τ_n a threshold, which can be calibrated a priori given some knowledge the tolerance to approximate monotonicity. However, such a knowledge

7 Adaptive Bayes Test for Monotonicity

may not be available; we thus propose an automatic calibration of τ_n such that our test has good asymptotic properties. To perform such a test we consider the $\gamma_0 - \gamma_1$ loss with fixed $\gamma_0, \gamma_1 > 0$ and thus our procedure can be defined as

$$\delta_n^\pi := \begin{cases} 0 \text{ if } \pi\left(\tilde{d}(f, \mathscr{F}) \leq \tau_n | X_n\right) \geq \frac{\gamma_0}{\gamma_0 + \gamma_1} \\ 1 \text{ otherwise.} \end{cases} \qquad (4)$$

This idea is similar to the one proposed in [16] and to the approximation of a point null hypothesis by an interval hypothesis testing; see also [20]. This test is straightforward to implement and will only require sampling under the posterior.

7.2 Theoretical Results

The following theorem gives a calibration for τ_n^k. It also gives an upper bound for the minimal separation rate with respect to the distance $d_n(\cdot, \cdot)$ defined as

$$d_n^2(f, g) = n^{-1} \sum_{i=0}^{n-1} \left(f\left(\frac{i}{n}\right) - g\left(\frac{i}{n}\right)\right)^2.$$

We define a prior π on f, σ similarly to before by considering

$$f_{\omega,k}(\cdot) = \sum_{i=1}^{k} \omega_i \mathbb{I}[(i-1)/k, i/k)(\cdot)$$

and

$$d\pi(\omega, \sigma, k) = \pi_k(k)\pi_\sigma(\sigma) \prod_{i=1}^{k} g(\omega_i),$$

where g and π_σ are density function. We consider the following conditions on the prior

- C1 The densities g and π_σ are continuous, $g(x) > 0$ for all $x \in \mathbb{R}$ and $\pi_\sigma(\sigma) > 0$ for all $\sigma \in (0, \infty)$.
- C2 π_k is such that there exists positive constants C_d and C_u such that

$$e^{-C_d k L(k)} \leq \pi_k(k) \leq e^{-C_u k L(k)}, \qquad (5)$$

where $L(k)$ is either $\log(k)$ or 1.

The condition C1 is mild as it is satisfied for a large variety of distributions. C2 is a usual condition when considering mixture models with random number of components (see, e.g., [17]) and is satisfied by Poisson or Geometric distribution, for instance. Similarly to what is done in the frequentist literature, we consider Hölderian alternatives $\mathscr{H}(\alpha, L)$ with $0 < \alpha \leq 1$ defined as

$$\mathscr{H}(\alpha, L) = \{f, \forall x, y \in [0,1], |f(y) - f(x)| \leq L|y-x|^\alpha\}.$$

Under the previous conditions, Theorem 1 gives us some insight on how to choose τ_n^k.

Theorem 1. *Under the assumptions C1 and C2, if $M_0 > 0$, setting $\tau_n^k = M_0\sqrt{k\log(n)/n}$ and δ_n^π the testing procedure*

$$\delta_n^\pi = \mathbb{I}\{\pi\left(H(\omega, k) > \tau_n^k | Y^n\right) > \gamma_0/(\gamma_0 + \gamma_1)\}$$

then there exists some $M > 0$ such that for all $\alpha \in (0, 1]$

$$\sup_{f \in \mathscr{F}} \mathrm{E}_f^n(\delta_n^\pi) = o(1)$$
$$\sup_{f, d_n(f, \mathscr{F}) > \rho, f \in \mathscr{H}(\alpha, L)} \mathrm{E}_f^n(1 - \delta_n^\pi) = o(1) \qquad (6)$$

for all $\rho > \rho_n = M(n/\log(n))^{-\alpha/(2\alpha+1)}\sqrt{\log(n)/L(n)}$.

Note that neither the prior nor the hyperparameters depend on the regularity α of the regression function under the alternative. Moreover, for all $\alpha \in (0, 1]$, the separation rate $\rho_n(\alpha)$ is the minimax separation rate up to a $\log(n)$ term. Thus our test is almost minimax adaptive. The $\log(n)$ term seems to follow from our definition of the consistency where we do not fix a level for the type I or type II error contrariwise to the frequentist procedures. The conditions on the prior are quite loose and are satisfied in a wide variety of cases. The constant M_0 does not influence the asymptotic behavior of our test but has a great influence in practice for finite n. However, it could easily be calibrated and our procedure leads to good results in practice.

7.3 Conclusion

We propose a Bayesian testing procedure that is consistent and is minimax adaptive up to a $\log(n)$ term. Furthermore our procedure is easy to implement and yield good results in the finite sample case. This is a great advantage as the existing frequentist procedures require in general heavy computations. Our approach can easily be adapted to testing other types of shape constrains such as convexity for instance.

References

1. Akakpo N, Balabdaoui F, Durot C (2012) Testing monotonicity via local least concave majorants. Bernoulli (preprint)
2. Baraud Y, Huet S, Laurent B (2005) Testing convex hypotheses on the mean of a Gaussian vector. Application to testing qualitative hypotheses on a regression function. Ann Stat 33(1):214–257
3. Barlow RE, Bartholomew DJ, Bremner JM, Brunk HD (1972) Statistical inference under order restrictions. The theory and application of isotonic regression. Wiley series in probability and mathematical statistics. Wiley, New York
4. Bornkamp B, Ickstadt K (2009) Bayesian nonparametric estimation of continuous monotone functions with applications to dose-response analysis. Biometrics 65(1):198–205
5. Brunner LJ, Lo AY (1989) Bayes methods for a symmetric unimodal density and its mode. Ann Stat 17(4):1550–1566
6. Ghosal S, Sen A, van der Vaart AW (2000) Testing monotonicity of regression. Ann Stat 28(4):1054–1082
7. Groeneboom P (1985) Estimating a monotone density. In: Proceedings of the Berkeley conference in honor of Jerzy Neyman and Jack Kiefer, vol II (Berkeley, California, 1983). Wadsworth statistical/probability series. Wadsworth, Belmont, CA, pp 539–555
8. Hall P, Heckman NE (2000) Testing for monotonicity of a regression mean by calibrating for linear functions. Ann Stat 28(1):20–39
9. Hussian M, Grimvall A, Burdakov O, and Sysoev O (2004) Monotonic regression for trend assessment of environmental quality data. In: The Proceedings of the 4th European Congress of Computational Methods in Applied Science and Engineering 'ECCOMAS 2004'
10. Khazaei S, Rousseau J, Balabdaoui F (2012) Nonparametric bayesian estimation of densities under monotonicity constraint.
11. Lo AY (1984) On a class of Bayesian nonparametric estimates: I. density estimates. Ann Stat 12:351–357
12. Mukerjee H (1988) Monotone nonparametric regression. Ann Stat 16(2):741–750
13. Neittaanmäki P, Rossi T, Majava K, Pironneau O (2008) Monotonic regression for trend assessment of environmental quality data. In: Hussian M, Grimvall A, Burdakov O, Sysoev O (eds) (2004) The proceedings of the 4th European congress of computational methods in applied science and engineering 'ECCOMAS 2004'
14. Prakasa Rao BLS (1970) Estimation for distributions with monotone failure rate. Ann Math Stat 41:507–519
15. Robertson T, Wright FT, Dykstra RL (1988) Order restricted statistical inference. Wiley series in probability and mathematical statistics: probability and mathematical statistics. Wiley, Chichester
16. Rousseau J (2007) Approximating interval hypothesis: p-values and Bayes factors. In: Bayesian statistics, vol 8, Oxford Science Publishing. Oxford University Press, Oxford, pp 417–452
17. Rousseau J (2010) Rates of convergence for the posterior distributions of mixtures of betas and adaptive nonparametric estimation of the density. Ann Stat 38(1):146–180
18. Salomond J-B (2013) Concentration rate and consistency of the posterior under monotonicity constraints. ArXiv e-prints, Jan 2013
19. Scott JG, Shively TS, Walker SG (2013) Nonparametric Bayesian testing for monotonicity. ArXiv e-prints, Apr 2013
20. Verdinelli I, Wasserman L (1998) Bayesian goodness-of-fit testing using infinite-dimensional exponential families. Ann Stat 26(4):1215–1241

Chapter 8
Bayesian Inference on Individual-Based Models by Controlling the Random Inputs

Michael Spence and Paul Blackwell

Abstract Complex models are becoming increasingly popular in ecological modelling. However, quantifying uncertainty, estimating parameters and so on for a model of this sort are complicated by the fact that their probabilistic behaviour is often implicit in its rules or programs rather than made explicit as in a more conventional statistical or stochastic model.

In a complex stochastic model, the output is dependent on both the parameters and the random inputs, i.e. the random numbers used to resolve decisions or generate stochastic quantities within the model. By treating these random inputs as nuisance parameters, often we can turn the model into a deterministic model where small movements in the parameter space result in small changes in the model output. When this is the case it will allow us to use Approximate Bayesian Computation methods with MCMC in order to perform parameter estimation. Controlling the random inputs allows us to move better in the parameter space and improves the mixing of the Markov chain.

We will use these methods to estimate parameters in an individual-based model which is used to model the population dynamics of a group-living bird, the woodhoopoe.

8.1 Introduction

In ecology the need for answering the question "what makes something happen?" as opposed to "what actually happens?" is becoming increasingly popular. These questions lead to building complex models where the different aspects of the system

M. Spence (✉) • P. Blackwell
School of Mathematics and Statistics, University of Sheffield, Sheffield, UK
e-mail: M.A.Spence@sheffield.ac.uk; p.blackwell@sheffield.ac.uk

are modelled separately and give rise to the collective behaviour of the system. A natural approach in ecology is to model each individual separately in the system. These are called individual-based models (IBMs) [5].

IBMs generally model behaviour through a series of rules or algorithms rather than describing it in a formal mathematical way. They are developed with algorithms that are not well tuned from the beginning but require parameters that are either not precise enough in the literature or simply not concretely measurable [6]. As the probabilistic behaviour of the model is implicit in the rules of the model, the likelihood is generally intractable.

Using pattern-oriented methods [7], parameter values can be found indirectly by changing a number of parameters at once and seeing if the model output matches some observed data [9, 10]. As IBMs have many parameters that can have complex and interacting effects on the output, this approach may be unproductive [2] so other methods of parameter estimation are required.

Current methods for performing parameter estimation for a model of this kind involve tuning the parameters and trying to match the model output to observed patterns. A range of potential parameter values are proposed and then tested by dividing the potential parameter space up and seeing if the output matches the observed data.

A more probabilistic version of this is ABC. In ABC parameter values are proposed from a distribution, usually the prior, and the model is run using these parameters. If the model output is similar to the realized data, then that parameter is accepted. This is continued until n parameter values have been accepted. Although this is more probabilistic it can be computationally very inefficient if the prior is not very informative.

We propose a method of examining the model by controlling the random inputs in the model which will allow us to turn the stochastic model into a deterministic one. We use this method, coupled with Approximate Bayesian Computation (ABC) methods, to perform parameter estimates on these types of models.

8.2 Controlling Random Inputs

Given a stochastic model M with input parameters θ and output

$$X \sim M(\cdot|\theta), \tag{1}$$

X is drawn from a random distribution. However, if we condition on the parts that cause the stochasticity, the random inputs u, then

$$X \sim M(\cdot|\theta, u) \tag{2}$$

will be deterministic.

By controlling the random inputs we aim to ensure that small changes in the parameters result in small changes in the model output. However, depending on how the inputs are used, a change in a parameter may cause a submodel to require a different number of random inputs to what it required before that could result in a large change in the output.

For example, consider a model where only the annual numbers of births and the weights at birth are generated. Let parameter θ control the number of births and u_i, $(i = 1, 2, \ldots)$ be the sequence of random inputs. Suppose when $\theta = \alpha$, using u_1, there is only one birth. The weight of the birth is determined by u_2. The next time the number of births will be determined using u_3. Now suppose that when $\theta = \alpha + \epsilon$, where ϵ is small, using u_1, there may now be two births whose weights involve u_2 and u_3. The next time the number of births will be determined using u_4. This causes all of the random inputs to be out of sequence which could result in a large change in the output. This change in output is caused by the change in sequence of random inputs not the parameter θ.

One way to get round this is to control all of the inputs individually so that each submodel will have its own sequence of inputs and then the inputs for each submodel will remain the same regardless to what happened in earlier submodels.

Another example is the queuing model described by Fearnhead and Prangle [3], Blum and François [1] and Heggland and Frigessi [4]. In this model the customers arrive at intervals determined by an exponential distribution with parameter θ_3 and are served one at a time with the service time being sampled from a uniform distribution on the interval $[\theta_1, \theta_1 + \theta_2]$. The output of the model is the inter-service times for the first 50 customers.

The stochastic inputs determine the arrival time and the service time denoted u and w, respectively, with

$$u_i \sim U(\cdot | 0, 1)$$

and

$$w_j \sim U(\cdot | 0, 1)$$

for $i = 1, 2, \ldots$ and $j = 1, \ldots, 50$. The ith arrival will be

$$-\frac{\log(1 - u_i)}{\theta_3}$$

after the $(i - 1)$th arrival and the jth service time will be

$$\theta_1 + \theta_2 w_j.$$

Given u and w, the model is then deterministic.

8.3 Woodhoopoe Model

Woodhoopoes are birds that can be found in sub-Saharan Africa [7]. They live in groups just like wolves, with one dominant pair which are the only ones that breed. Neuert et al. [8] used an individual-based model in order to model the population and group dynamics of the woodhoopoes which was simplified by Railsback and Grimm [7] for use as examples in their recent book.

In the model individual woodhoopoes live in groups with one dominant male and one dominant female. Each woodhoopoe's aim is to become a dominant animal in one of the groups. Each month, each woodhoopoe dies with a probability of θ_1. When a dominant dies the eldest subordinate, if there is any, will become the dominant animal. Younger subordinates will leave their group in order to try and become a dominant in another group with probability θ_2, but if they leave the safety of the group, they become more vulnerable to predators and have a predication mortality probability of θ_3.

8.4 Summary of the Talk

In the talk we will introduce this method of controlling the random inputs, show a few examples of how to do it on toy models and then, along with ABC methods, use it to perform parameter estimation on Railsback and Grimm's [7] woodhoopoe model.

References

1. Blum MGB, François O (2010) Non-linear regression models for approximate Bayesian computation. Stat Comput 20:63–73
2. Grimm V, Railsback S (2005) Individual-based modelling and ecology. Princeton series in theoretical and computational biology. Princeton University Press, Princeton
3. Fearnhead P, Prangle D (2012) Constructing summary statistics for approximate Bayesian computation: semi-automatic approximate Bayesian computation. J Roy Stat Soc 74:1–28
4. Heggland K, Frigessi A (2002) Estimating functions in indirect inference. J Roy Stat Soc 66:447–462
5. Kaiser H (1979) The dynamics of populations as a result of the properties of individual animals. Fortschritte der Zoologie 25:109–136
6. Piou C, Berger V, Grimm V (2009) Proposing an information criterion for individual-based models developed in a pattern-oriented modelling framework. Ecol Modelling 220:1957–1967
7. Railsback S, Grimm V (2012) Agent-based and individual-based modeling a practical introduction. Princeton University Press, Princeton
8. Neuert C, du Plessis M, Grimm V, Wissel C (1995) Welche ökologischen Faktoren bestimmen die Gruppengröße bei Phoeniculus purpureus (Gemeiner Baumhopf) in Südafrika? Ein individuenbasierte Modell. Verhandlungen der Gesellschaft für Ökologie 24:145–149

9. Wiegand T, Jeltsch F, Hanski I, Grimm V (2003) Using pattern-oriented modeling for revealing hidden information: a key for recording ecological theory and application. Oikos 100:209–222
10. Wiegand T, Revilla E, Knauer F (2004) Dealing with uncertainty in spatial explicit population models. Biodivers Conv 13:53–78

Chapter 9
Consistency of Bayesian Nonparametric Hidden Markov Models

Elodie Vernet

Abstract We are interested in Bayesian nonparametric hidden Markov models. More precisely, we are going to present the consistency of these models under appropriate conditions on the prior distribution, when the number of states of the Markov chain is finite and known. Our approach is based on exponential forgetting and usual Bayesian consistency techniques.

9.1 Introduction

Hidden Markov models are much used in practice as in econometrics, speech recognition, genomics (see [2] for some applications), etc. Since the 1960s, algorithms to estimate the parameters or the hidden states of these models have been developed without many theoretical results. Since the 1990s, statisticians have studied the frequentist asymptotic properties of hidden Markov model. Namely they have proved consistency and then the asymptotic normality of the maximum likelihood estimator in general parametric cases [2]. In the nonparametric case the problem of identifiability is not trivial. It has been solved recently in [5] in the case we are going to study here. The authors also exhibit frequentist estimators which have good asymptotic and non-asymptotic properties in the semi-parametric and nonparametric case. As to the Bayesian asymptotic results, there were recently studied and only in the parametric case [6, 7]. To our knowledge, there is no result in the Bayesian nonparametric case. Yet Yau et al. in [8] noticed that using a nonparametric model can improve the estimations a lot. In this paper we will prove that under some assumptions, the posterior is consistent for some nonparametric Bayesian hidden Markov model.

E. Vernet (✉)
Laboratoire de Mathématiques, Université Paris-Sud, Orsay, France
e-mail: elodie.vernet@math.u-psud.fr

Fig. 9.1 The model

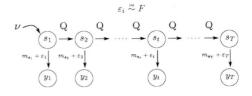

9.2 The Model

We are interested in hidden Markov models. Let $S_1, \ldots, S_T$ be a Markov chain with a finite and known number of states k, an initial probability ν, and transition matrix $Q \in \mathcal{M}_{k,k}$.

With hidden Markov chain, we cannot access these states (they are hidden). But we observe $Y_1, \ldots, Y_T$ which are the noisy signals of the states of the chain, given $S_1, \ldots, S_T$; $Y_1, \ldots, Y_T$ are independent and Y_t is equal to a parameter m_{S_t} depending on the corresponding state S_t plus a noise ϵ_t. We assume that $\epsilon_1, \ldots, \epsilon_T$ are iid, distributed according to a probability F and independent of the Markov chain (Fig. 9.1).

The parameters of this model are ν, Q, m and F. We assume that F has a density f with respect to a reference measure λ. Let $\theta = (\nu, Q, m, f)$ and p_θ^T be the associated joint density of $Y_1, \ldots Y_T$ with respect to λ.

As usual in Bayesian statistics, we put a prior μ on the parameters. We assume that this prior is a product of probability measure of each parameter sets. Then we take a "frequentist point of view" by wondering if the posterior puts asymptotically the mass on the neighborhood of the true density. In other words, we study the consistency of the posterior.

9.3 Consistency

Consistency is the first thing we may ask. Here we will work with neighborhood with respect to the l_1 distance between two joint densities of order l. For two parameters θ and θ' the pseudo-metric between the two of them will be $\int |p_\theta(y_1, \ldots, y_l) - p_{\theta'}(y_1, \ldots, y_l)| \lambda(dy_1) \ldots \lambda(dy_l)$.

Theorem 1. *Under some assumptions on the set of parameters and the prior, if the true parameter $\theta^* = (\nu^*, Q^*, m^*, f^*)$ is such that the associated Markov chain mixes enough then the posterior is consistent with respect to the previously described pseudo-metric.*

The assumptions on the set of parameters are usual. We mostly ask that the Markov chain mixes enough. The assumptions on the prior consist on trivial assumptions on the prior on the parametric part and more complex ones on the nonparametric part. These last assumptions are checked for some mixtures of Gaussians on f.

This result is proved using Barron method [1]. That is to say we have to prove that the parameter set is not too big by proving the existence of tests. This task can be achieved by the construction made in [6]. Secondly, we have to prove that the posterior puts enough mass in the Kullback–Leibler neighborhood of the true density. For this purpose, we need to control a nasty Kullback–Leibler divergence by controlling the parameters. We did it thanks to existing results on hidden Markov chains [3, 4].

Acknowledgements I want to thank Elisabeth Gassiat and Judith Rousseau for their precious help.

References

1. Barron AR (1988) The exponential convergence of posterior probabilities with implications for Bayes estimators of density functions. Technical report, Apr 1988
2. Cappé O, Moulines E, Rydén T (2005) Inference in hidden Markov models. Springer, New York
3. Douc R, Matias C (2001) Asymptotics of the maximum likelihood estimator for general hidden Markov models. Bernouilli 7:381–420
4. Douc R, Moulines E, Rydén T (2004) Asymptotic properties of the maximum likelihood estimator in autoregressive models with Markov regime. Ann Stat 32(5):2254–2304
5. Gassiat E, Rousseau J (2013) Non parametric finite translation mixtures with dependent regime (submitted)
6. Gassiat E, Rousseau J (2012) About the posterior distribution in hidden markov models with unknown number of states. Bernouilli (to appear)
7. de Gunst MC, Shcherbakova O (2008) Asymptotic behavior of Bayes estimators for hidden Markov models with application to ion channels. Math Methods Stat 17(4):342–356
8. Yau C, Papaspiliopoulos O, Roberts GO, Holmes C (2011) Bayesian non-parametric hidden Markov models with applications in genomics. J Roy Stat Soc 73:37–57

Chapter 10
Bayesian Methodology in the Stochastic Event Reconstruction Problems

Anna Wawrzynczak, Piotr Kopka, and Mieczyslaw Borysiewicz

Abstract In many areas of application it is important to estimate unknown model parameters in order to model precisely the underlying dynamics of a physical system. In this context the Bayesian approach is a powerful tool to combine observed data along with prior knowledge to gain a current (probabilistic) understanding of unknown model parameters. We have applied the methodology combining Bayesian inference with sequential Monte Carlo (SMC) to the problem of the atmospheric contaminant source localization. The algorithm input data are the on-line arriving information about concentration of given substance registered by the downwind distributed sensor's network. We have proposed the different version of the hybrid SMC along with Markov chain Monte Carlo (MCMC) algorithms and examined its effectiveness to estimate the probabilistic distributions of atmospheric release parameters.

10.1 Introduction

Accidental atmospheric releases of hazardous material pose great risks to human health and the environment. Examples like Chernobyl nuclear power plant accident in 1986 in Ukraine, chemical plants producing or storing dangerous materials (e.g. Seveso disaster in 1978) or transportation accidents (bromine release on the train in Chelyabinsk in 2011) prove that it is necessary to have properly fast response to such incidents. In this context it is valuable to develop the emergency action support system, which based on the concentration measurement of dangerous substance by the network of sensors can identify probable location and characteristics of the release source.

A. Wawrzynczak (✉) • P. Kopka • M. Borysiewicz
National Centre for Nuclear Research, ul. Andrzeja Sołtana 7, 05-400 Świerk-Otwock, Poland
e-mail: a.wawrzynczak@ncbj.gov.pl; piotr.kopka@ncbj.gov.pl; manhaz@ncbj.gov.pl

It is obvious that if we are able to create the model giving the same point concentration of registered substance, as we get from the sensors' network, we could say that we understand the situation we face up. However, to create the model realistically reflecting the real situation based only on a sparse point-concentration data is not trivial. This task requires specification of set of models' parameters, which depends on the applied dispersion model's characteristics.

In general, the stated inverse problem for the dispersion of released materials in the air is ill-posed. Given concentration measurements and knowledge of the wind field and other atmospheric air parameters, finding the location of the source and its parameters is ambiguous. This problem has no unique solution and can be considered only in the probabilistic frameworks. In the case of gas dispersion, the unknowns to be determined are the gas source distribution of strengths and locations, given the measured gas concentrations at measurement locations for the associated wind field and other weather data (e.g. weather stability pattern). In fact, our aim is to find the source parameter's distributions that will generate predicted concentrations closest to those actually measured.

In this paper we present the developed stochastic dynamic data-driven event reconstruction model which couples data and predictive models through Bayesian inference to obtain a solution to the inverse problem, i.e. based on the successively arriving information about concentration of given substance registered by distributed sensor network find the most probable source location and its strength.

10.2 Theoretical Preliminaries

Bayes' theorem, as applied to an emergency release problem, can be stated as follows:

$$P(M|D) \propto P(D|M)P(M), \qquad (1)$$

where M represents possible model configurations or parameters and D are observed data. For our problem, Bayes' theorem describes the conditional probability $P(M|D)$ of certain source parameters (model configuration M) given observed measurements of concentration at sensor locations (D). This conditional probability $P(M|D)$ is also known as the posterior distribution and is related to the probability of the data conforming to a given model configuration $P(D|M)$ and to the possible model configurations $P(M)$, before taking into account the sensors' measurements. The probability $P(D|M)$, for fixed D, is called the likelihood function, while $P(M)$ is the prior distribution [3].

Value of likelihood for a sample is computed by running a forward dispersion model with the given source parameters M. To achieve the rapid-response event reconstructions and limit the computational time we have adopted the fast-running Gaussian plume dispersion model [4] as the forward dispersion model. The model predicted concentrations M in the points of sensor location are compared with actual data D. The closer the predicted values are to the measured ones, the higher is the likelihood of the sampled source parameters. This function is taken as

$$\ln[P(D|M)] = \ln[\lambda(M)] = -\frac{\sum_{i=1}^{N}[\log(C_i^M) - \log(C_i^E)]^2}{2\sigma_{\text{rel}}^2} \qquad (2)$$

where λ is the likelihood function, C_i^M are the predicted by the forward model concentrations at the sensor locations i, C_i^E are the sensor measurements, σ_{rel}^2 is the standard deviation of the combined forward model and measurement errors and N is the number of sensors.

We use a sampling procedure with the Metropolis-Hastings algorithm to obtain the posterior distribution $P(M|D)$ of the source term parameters given the concentration measurements at sensor locations. This way we completely replace the Bayesian formulation with a stochastic sampling procedure to explore the model parameters' space and to obtain a probability distribution for the source location [1,2]. The scanned model's parameter space is $M \equiv M(x, y, q, \zeta_1, \zeta_2)$ where x and y are the spatial location of the release, q is release rate and ζ_1, ζ_2 are stochastic terms in the turbulent diffusion parameters.

10.3 Methods and Results

In this paper we examine the application of the sequential Monte Carlo (SMC) methods combined with the Bayesian inference to the problem of the localization of the atmospheric contamination source. We present the possibility to connect MCMC and SMC to provide additional benefit in the process of event reconstruction. Based on the synthetic release experiment we have proposed and tested the following version of the hybrid SMC with MCMC algorithms in effectiveness to estimate the probabilistic distributions of searched parameters:

1. **Classic MCMC.** In this algorithm, the parameter space scan in each time step t is independent from the previous ones. So, in this case we do not use information from past calculations. Classic MCMC algorithms do not use sequential mechanism.
2. **MCMC Prior to SMC.** The SMC algorithm uses the set of samples generated by K iterations of classic MCMC algorithm as a prior distribution, but in subsequent SMC iterations do not use information from SMC results from previous time step.
3. **SMC Via Maximal Weights.** In subsequent SMC calculations algorithms use the results obtained by SMC in the previous time steps to run calculation with the use of the new measurements. As the first location of Markov chain M_0^t it selects the set of M parameters for which weight in previous time step was the highest. So, for $t > 1$:

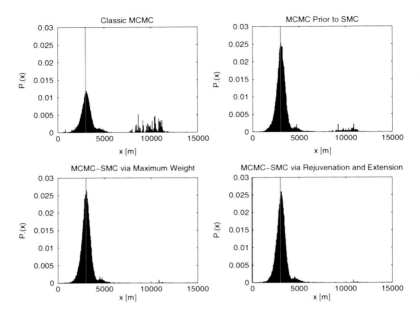

Fig. 10.1 Posterior distribution as inferred by the Bayesian event reconstruction for all applied algorithms for x parameter. Vertical lines represent the target x value

$M_0^t \sim \arg(M \in \{M_0^{t-1}, \ldots, M_n^{t-1}\}) \max w(M_i^{t-1})$. With this approach, we always start with the best values of the model (previously found) and correct the result with new information from sensor.

4. **SMC Via Rejuvenation and Extension.** In contrast to the SMC via maximal weights this algorithm as the first location of Markov chain M_0^t at the time $t > 1$ chooses the set of parameters M selected randomly from previous realization of resampling procedure in $t - 1$ with use of the uniform distribution: $M_0^t \sim U(M_0^{t-1}, M_1^{t-1}, \ldots, M_n^{t-1})$ a uniform distribution $\{1, \ldots, n\}$. Applying the new knowledge (new measurements) the current chain is "extended" starting from selected position with use of the new data in the likelihood function calculation.

We have shown the advantage of the algorithms that in different ways use the source location parameter's probability distributions obtained basing on available measurements to update the marginal probability distribution. As the most effective we pointed the modifications of MCMC prior to SMC (see Fig. 10.1).

Acknowledgements This work was supported by the Welcome Programme of the Foundation for Polish Science operated within the European Union Innovative Economy Operational Programme 2007–2013 and by the EU and MSHE grant nr POIG.02.03.00-00-013/09.

References

1. Borysiewicz M, Wawrzynczak A, Kopka P (2012) Bayesian-based methods for the estimation of the unknown model's parameters in the case of the localization of the atmospheric contamination source. Found Comput Decis Sci 37(4):253–270
2. Borysiewicz M, Wawrzynczak A, Kopka P (2012) Stochastic algorithm for estimation of the model's unknown parameters via Bayesian inference. Proceedings of the federated conference on computer science and information systems, Wroclaw, pp 501–508
3. Gilks W, Richardson S, Spiegelhalter D (1996) Markov chain Monte Carlo in practice. Chapman Hall/CRC, Boca Raton
4. Panofsky HA, Dutton JA (1984) Atmospheric turbulence. Wiley, New York

Part II
Computational Bayes

Chapter 11
Efficient Fitting of Bayesian Regression Models with Spatio-Temporally Varying Coefficients

Mark Bass and Sujit Sahu

Abstract Bayesian regression models with spatio-temporally varying coefficients are gaining popularity among researchers who are looking to model the spatio-temporal processes that are ubiquitous in the environmental and physical sciences. The fitting of these highly overparameterised and non-stationary models is challenging and computationally expensive. By combining existing ideas of reparameterisation, marginalisation and interweaving we develop a number of hybrid fitting strategies. We use the MCMC output to compare these methods in terms of convergence rates and effective sample sizes per second and thus identify the most efficient fitting strategy for models of this type. Implementation of the optimal strategy achieves faster convergence rates and significant savings in computation time, illustrated here with a simulation example and also a real data example modelling daily ozone concentration data.

11.1 Introduction

Given that large data sets are now prevalent in many areas of statistics, it is important to efficiently implement any Markov chain Monte Carlo (MCMC) algorithm. It has long been understood that the parameterisation of a hierarchical model affects the performance of the MCMC method used for inference. In particular, high posterior correlations between model parameters can lead to poor mixing and slow convergence.

Papaspiliopoulos et al. [5] develop a framework for the parameterisation of hierarchical models applied to a range of statistical contexts. They focus on two parameterisations, namely *centred* and *noncentred*, terms introduced by Gelfand et al. [3]. Where a model can be parameterised in these two ways Yu and Meng [8] promote the use of an interviewing strategy to reduce Markovian dependence.

One might also consider marginalising over any latent variable, thus reducing the dimension of the posterior parameter space. This practice is adopted for the fitting of spatial models in the **R** package spBayes [2].

To illustrate these ideas consider the following simple hierarchical model [4]

$$Y = W + \sigma_y \epsilon,$$
$$W = \theta + \sigma_w \epsilon, \quad (1)$$

where Y is the observed datum, W is a latent variable and θ is the parameter of interest. Independent errors ϵ follow a standard normal distribution and σ_y and σ_w are assumed to be known. We refer to this parameterisation as centred, Y is centred on W and W is centred on θ. Assuming an improper uniform prior for θ, we get the following conditional posterior distributions for W and θ:

$$W|\theta, y \sim N\left(\frac{\sigma_y^2 \theta + \sigma_w^2 y}{\sigma_y^2 + \sigma_w^2}, \frac{\sigma_y^2 \sigma_w^2}{\sigma_y^2 + \sigma_w^2}\right), \quad (2)$$

$$\theta|W, y \sim N(W, \sigma_w^2). \quad (3)$$

A standard algorithm iterates between (2) and (3) by drawing $W^{(t)} \sim \pi(W|\theta^{(t)}, y)$ and then drawing $\theta^{(t+1)} \sim \pi(\theta|W^{(t)}, y)$.

If we consider the random variable $\tilde{W} = W - \theta$, then we can rewrite model (1) in its noncentred form:

$$Y = \tilde{W} + \theta + \sigma_y \epsilon,$$
$$\tilde{W} = \sigma_w \epsilon. \quad (4)$$

Under this parameterisation the conditional posterior distributions for $\tilde{W}$ and θ are

$$\tilde{W}|\theta, y \sim N\left(\frac{\sigma_w^2(y-\theta)}{\sigma_y^2 + \sigma_w^2}, \frac{\sigma_y^2 \sigma_w^2}{\sigma_y^2 + \sigma_w^2}\right) \quad (5)$$

$$\theta|\tilde{W}, y \sim N(y - \tilde{W}, \sigma_w^2). \quad (6)$$

Using the noncentred parameterisation we iterate between (5) and (6). Therefore, each parameterisation gives us a different algorithm and although they have the same target distribution, $\pi(\theta|y)$, they typically have different convergence rates.

Where the joint distribution of $\pi(W, \tilde{W}|\theta, y)$ is well defined we can implement the interweaving algorithm [8]. Given $\theta^{(t)}$, we obtain a value for $\theta^{(t+1)}$ as follows:

Step 1. Draw $W^{(t)} \sim \pi(W|\theta^{(t)}, y)$.
Step 2. Draw $\theta^{(t+0.5)} \sim \pi(\theta|W^{(t)}, y)$.
Step $\tilde{2}$. Draw $\tilde{W}^{(t+1)} \sim \pi(\tilde{W}|\theta^{(t+0.5)}, W^{(t)}, y)$.
Step 3. Draw $\theta^{(t+1)} \sim \pi(\theta|\tilde{W}^{(t+1)}, y)$,

where $\theta^{(t+0.5)}$ is an intermediate draw which can be discarded. Often, W and $\tilde{W}$ are related via some deterministic function and step $\tilde{2}$ requires a trivial transformation. In this example, $\tilde{W}^{(t+1)} = W^{(t)} - \theta^{(t+0.5)}$.

However, we need not sample from W at all if it is integrated out of the likelihood. For both the centred and noncentred parameterisations we get a marginalised likelihood for Y as $Y|\theta \sim N(\theta, \sigma_y^2 + \sigma_w^2)$, and thus we get a marginal posterior distribution for θ as $\theta|y \sim N(y, \sigma_y^2 + \sigma_w^2)$.

In this paper we aim to develop the most efficient way to implement a non-linear model which involves spatio-temporally correlated latent stochastic processes. We investigate how the ideas of reparameterisation, marginalisation and interweaving can be used to develop an efficient fitting strategy.

11.2 A Spatio-Temporal Model

Let $Y(\mathbf{s}, t)$ denote an observation at site $\mathbf{s}$ and at time t. Further, let $x(\mathbf{s}, t)$ be the value of a single predictor for Y, again at site $\mathbf{s}$ and at time t. Consider the following spatio-temporal model [1] written in its noncentred form:

$$Y(\mathbf{s}_i, t) = \beta_0 + \tilde{\beta}_0(\mathbf{s}_i, t) + \{\beta_1 + \tilde{\beta}_1(\mathbf{s}_i, t)\}x(\mathbf{s}_i, t) + \epsilon(\mathbf{s}_i, t), \quad (7)$$

for i=1,...,n, t=1,...,T, where $\epsilon(\mathbf{s}_i, t) \stackrel{ind}{\sim} N(0, \sigma_\epsilon^2)$. We take β_0 to be the fixed intercept and β_1 as the fixed regression coefficient. These are locally perturbed by $\tilde{\beta}_0(\mathbf{s}_i, t)$ and $\tilde{\beta}_1(\mathbf{s}_i, t)$, respectively, which are both modelled as zero mean Gaussian processes. Note that we could consider model (7) to be a multivariate spatial model where data is collected for T variables at each site $\mathbf{s}$.

Denote by $\mathbf{x}$ the vector containing all nT values of $x(\mathbf{s}_i, t)$ and let $\mathbf{D}_x = \text{diag}(\mathbf{x})$; then with matrix $\tilde{\mathbf{X}} = (\mathbf{1}, \mathbf{I}, \mathbf{x}, \mathbf{D}_x)$, we can write the likelihood as

$$\mathbf{Y} \sim N(\tilde{\mathbf{X}}\tilde{\boldsymbol{\beta}}, \boldsymbol{\Sigma}), \quad (8)$$

where $\boldsymbol{\Sigma} = \sigma_\epsilon^2 \mathbf{I}$. We represent by $\mathbf{I}$ the $nT \times nT$ identity matrix and let $\tilde{\boldsymbol{\beta}} = (\beta_0, \tilde{\boldsymbol{\beta}}_0, \beta_1, \tilde{\boldsymbol{\beta}}_1)'$, where $\tilde{\boldsymbol{\beta}}_k = (\tilde{\beta}_k(\mathbf{s}_1, 1), \ldots, \tilde{\beta}_k(\mathbf{s}_n, T))' \sim N(0, \boldsymbol{\Sigma}_k)$, $k = 0, 1$.

11.2.1 Parameterisation, Marginalisation and Interweaving

A different parameterisation of the model can be found by centering the spatio-temporal processes on the fixed parameters in the mean. Let $\beta_k(\mathbf{s}, t) = \tilde{\beta}_k(\mathbf{s}, t) + \beta_k$, $k = 0, 1$, then we can rewrite model (7) in its centred form to get

$$Y(\mathbf{s}_i, t) = \beta_0(\mathbf{s}_i, t) + \beta_1(\mathbf{s}_i, t) x(\mathbf{s}_i, t) + \epsilon(\mathbf{s}_i, t), \quad i = 1, \ldots, n, \quad t = 1, \ldots, T, \tag{9}$$

where $\beta_0(\mathbf{s}_i, t)$ has mean β_0 and $\beta_1(\mathbf{s}_i, t)$ has mean β_1. For the centred parameterisation we have

$$\mathbf{Y} \sim N(\mathsf{X}\boldsymbol{\beta}, \boldsymbol{\Sigma}), \tag{10}$$

where $\mathsf{X} = (\mathsf{I}, \mathsf{D}_x)$ and $\boldsymbol{\beta} = (\boldsymbol{\beta}_0, \boldsymbol{\beta}_1)'$.

Each parameterisation of the model defines a fitting strategy and we will label these with respect to their likelihood functions. Where likelihood (8) is used the method will be referred to as *full noncentred* (Fnc) and where likelihood (10) is used we label the method *full centred* (Fc). Given these parameterisations, we can also use the interweaving algorithm to obtain a third fitting method (IF).

As the spatio-temporal processes are given Gaussian priors, we can easily integrate them out of the model. Taking the noncentred likelihood and marginalising over $\tilde{\boldsymbol{\beta}}_0$ gives

$$\mathbf{Y} \sim N(\beta_0 \mathbf{1} + \beta_1 \mathbf{x} + \mathsf{D}_x \tilde{\boldsymbol{\beta}}_1, \boldsymbol{\Sigma} + \boldsymbol{\Sigma}_0)$$

and achieves a reduction in the dimension of the parameter space of nT. If we fit the model in this way we label the method M0nc. Alternatively, if we reparameterise and then marginalise over $\boldsymbol{\beta}_0$, we have $\mathbf{Y} \sim N(\beta_0 \mathbf{1} + \mathsf{D}_x \boldsymbol{\beta}_1, \boldsymbol{\Sigma} + \boldsymbol{\Sigma}_0)$, and this method we label M0c. Further, we can interweave M0nc and M0c to obtain another fitting method, IM0. Similarly, we could have considered $\tilde{\boldsymbol{\beta}}_1$ and $\boldsymbol{\beta}_1$ to get fitting methods M1nc, M1c and IM1. Finally, the tenth strategy we consider is to marginalise over both $\tilde{\boldsymbol{\beta}}_0$ and $\tilde{\boldsymbol{\beta}}_1$ (or $\boldsymbol{\beta}_0$ and $\boldsymbol{\beta}_1$). Regardless of the parameterisation the resulting likelihood is

$$\mathbf{Y} \sim N(\beta_0 \mathbf{1} + \beta_1 \mathbf{x}, \boldsymbol{\Sigma} + \boldsymbol{\Sigma}_0 + \mathsf{D}_x \boldsymbol{\Sigma}_1 \mathsf{D}_x'),$$

which will be called M2.

By combinations of parameterisation, marginalisation and interweaving we obtain ten strategies for fitting the same model.

11.2.2 Model Specifications

A separable covariance structure is placed on the $\tilde{\beta}_k(\mathbf{s},t)$'s, given by

$$\text{Cov}\{\tilde{\beta}_k(\mathbf{s}_i,t_l),\tilde{\beta}_k(\mathbf{s}_j,t_m)\} = \sigma_k^2 \rho_s(|\mathbf{s}_i-\mathbf{s}_j|;\phi_k^s)\rho_t(|t_l-t_m|;\phi_k^t), \quad k=0,1,$$

where ρ_s and ρ_t are valid isotropic covariance functions from the Matérn family. Here, an exponential covariance functions is used, i.e., $\rho_s(d_s;\phi) = \exp(-\phi|d_s|)$ and $\rho_t(d_t;\phi) = \exp(-\phi|d_t|)$. Each of the ϕ is given a uniform prior. These parameters control the rates of decay of the spatial and temporal correlation between the random effects. Variance parameters σ_ϵ^2, σ_0^2 and σ_1^2 are modelled on their inverse scales and are given gamma priors, i.e., $1/\sigma_\epsilon^2 \sim Ga(a,b)$, where the gamma distribution has mean a/b. Fixed mean parameters β_0 and β_1 are given vague normal priors.

11.3 Results

As a simulation exercise, we generate data for T=7 time points at n=20 sites, randomly chosen over the unit square. The model is fitted by each of the ten strategies described in Sect. 11.2.1, running the chains for $N = 10,000$ iterations and discarding the first $M = 2,500$. Algorithms are written in the C programming language and performed on a single 2.4 Ghz Intel Westmere processor on the IRIDIS high performance computing facility. Run times (in seconds) for the Fnc, Fc and IF were 111. For the semi-marginalised methods M0nc, M0c and IM0 run times were 454 and for M1nc, M1c and IM1 they were 459. For the fully marginalised method the run time was 704. Clearly these times are machine and programmer dependent, but indicate the additional computing time required if marginalisation is considered.

The effective sample sizes for each parameter are computed in the **R** package coda [6], by dividing the number of samples, $N - M$, by an estimate of the autocorrelation time, $\kappa = 1 + 2\sum_{k=1}^{\infty} \rho(k)$, where $\rho(k)$ is the autocorrelation at lag k. For a fair comparison, effective sample sizes are divided by run times to give effective sample sizes per second (ESS/s). This is repeated for 100 data sets and we plot the average ESS/s for each model parameter, Fig. 11.1.

We see that IF gives the greatest ESS/s for β_0, σ_0^2 and σ_ϵ^2. Although M0nc and IM0 return the greatest ESS/s for β_1, they perform badly for σ_1^2, ϕ_1^s and ϕ_1^t.

One might consider M2 as a candidate for 'preferred fitting method', but it should be noted that with longer running times and variance and phi parameters all requiring metropolis updates, it is more difficult to tune than the other methods and more so with increasing nT. Hence, we promote the use of the IF method.

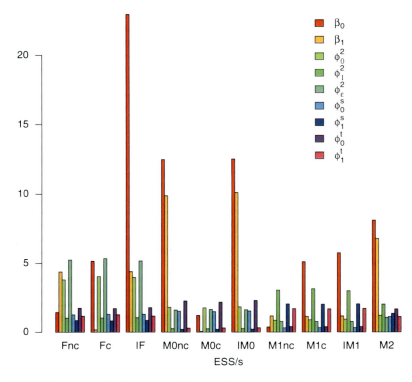

Fig. 11.1 Effective sample size per second for model parameters under each fitting method

11.4 Summary

By combining the procedures of reparameterisation, marginalisation and interweaving, we have developed ten hybrid strategies for fitting a spatio-temporal model. Although no one method returns the greatest ESS/s for all model parameters, the IF method has an advantage in implementation. Metropolis updates are only required for the decay parameters. A reduction in computing time is possible if one considers letting $\phi_0^s = \phi_1^s$ and $\phi_0^t = \phi_1^t$. In applications decay parameters are often fixed, see, e.g., [1,7]. If this modelling decision is made, then we have fully Gibbs scheme, and a further saving in run times is achieved.

References

1. Berrocal V, Gelfand A, Holland D (2010) A spatio-temporal downscaler for output from numerical models. J Agr Biol Environ Stat 15(2):176–197
2. Finley A, Banerjee S, Carlin B (2007) spBayes: an R package for univariate and multivariate hierarchical point-referenced models. J Stat Software 19(4):1–24

3. Gelfand A, Sahu S, Carlin B (1995) Efficient parameterisations for normal linear mixed models. Biometrika 82(3):479–488
4. Lui J, Wu Y (1999) Parameter expansion for data augmentation. J Am Stat Assoc 94:1264–1274
5. Papaspiliopoulos O, Roberts G, Sköld M (2007) A general framework for the parameterization of hierarchical models. Stat Sci 22(1):59–73
6. Plummer M, Best N, Cowles K, Vines K (2006) CODA: convergence diagnosis and output analysis for MCMC. R News 6(1):7–11
7. Sahu S, Yip S, Holland D (2011) A fast Bayesian method for updating and forecasting hourly ozone levels. Environ Ecol Stat 18:185–207
8. Yu Y, Meng X-L (2011) To center or not to center: that is not the question - an ancillarity-sufficiency interweaving strategy (ASIS) for boosting MCMC efficiency. J Comput Graph Stat 20(3):531–570

Chapter 12
PAWL-Forced Simulated Tempering

Luke Bornn

Abstract In this short note, we show how the parallel adaptive Wang–Landau (PAWL) algorithm of Bornn et al. (J Comput Graph Stat, to appear) can be used to automate and improve simulated tempering algorithms. While Wang–Landau and other stochastic approximation methods have frequently been applied within the simulated tempering framework, this note demonstrates through a simple example the additional improvements brought about by parallelization, adaptive proposals, and automated bin splitting.

12.1 A Parallel Adaptive Wang–Landau Algorithm

The central idea underlying Wang–Landau [6] and related algorithms is that instead of generating samples from a target density π, it is sometimes more efficient to instead sample a strategically biased density $\tilde{\pi}$. In the case of Wang–Landau, the goal is to sample

$$\tilde{\pi}(x) = \pi(x) \times \frac{1}{d} \sum_{i=1}^{d} \frac{\mathscr{I}_{\mathscr{X}_i}(x)}{\int_{\mathscr{X}_i} \pi(x) \mathrm{d}x}, \qquad (1)$$

where $\mathscr{I}_{\mathscr{X}_i}(x)$ is equal to 1 if $x \in \mathscr{X}_i$ and 0 otherwise. Interestingly, this biased target ensures each of the partitions of the space $(\mathscr{X}_i)_{i=1}^{d}$ are visited equally: $\int_{\mathscr{X}_i} \tilde{\pi}(x) \mathrm{d}x = \int_{\mathscr{X}_j} \tilde{\pi}(x) \mathrm{d}x$, $\forall i, j \in (1, \ldots, d)$. Additionally, the restriction of the modified distribution $\tilde{\pi}$ to each set $\mathscr{X}_i$ coincides with the restriction of the target distribution π to this set up to a multiplicative constant, namely for all i, $\tilde{\pi}(x) \propto \pi(x)$, $\forall x \in \mathscr{X}_i$.

L. Bornn (✉)
Harvard University, 1 Oxford St., Cambridge, MA 02138, USA
e-mail: bornn@stat.harvard.edu

While the biased density $\tilde{\pi}(x)$ has desirable properties, an obvious problem is that calculating $\int_{\mathcal{X}_i} \pi(x) \mathrm{d}x$ is not straightforward. As such, the Wang–Landau algorithm creates estimates θ_t of these quantities at each step t. Algorithm 1 provides psuedo-code for the algorithm. In the full version of the algorithm, the step

---
Algorithm 1 Simplified Wang–Landau algorithm

1: Partition the state space into d regions $\{\mathcal{X}_1, \ldots, \mathcal{X}_d\}$ along a reaction coordinate $\xi(x)$.
2: First, $\forall i \in \{1, \ldots, d\}$ set $\theta(i) \leftarrow 1$.
3: Choose a decreasing sequence $\{\gamma_t\}$, typically $\gamma_t = 1/t$.
4: Sample X_0 from an initial distribution π_0.
5: **for** $t = 1$ to T **do**
6:     Sample X_t from $P_{\theta_{t-1}}(X_{t-1}, \cdot)$, a transition kernel with invariant distribution $\tilde{\pi}_{\theta_{t-1}}(x)$.
7:     Update the bias: $\log \theta_t(i) \leftarrow \log \theta_{t-1}(i) + \gamma_t(\mathcal{I}_{\mathcal{X}_i}(X_t) - d^{-1})$.
8:     Normalize the bias: $\theta_t(i) \leftarrow \theta_t(i)/\sum_{i=1}^{d} \theta_t(i)$.
9: **end for**
---

size γ_t is only reduced when all of the regions $\{\mathcal{X}_1, \ldots, \mathcal{X}_d\}$ have been uniformly explored as measured by the flat histogram criterion $\max_{i \in [1,d]} |\nu(i) - d^{-1}| < c/d$ where $\nu(i)$ is the proportion of samples within $\mathcal{X}_i$ since the last time the flat histogram criterion was met. Here c is a user-specified threshold. The reader is referred to [3] for a full description and discussion of the algorithm, as well as details on stabilizing the algorithm through parallelization, introducing adaptive proposals, and automating the partitioning of the space. These three improvements, applied to simulated tempering, will be the focus of this work.

12.2 Simulated Tempering

The use of stochastic approximation algorithms, including Wang–Landau, within simulated tempering has been suggested by various authors (see, e.g., [1,4]). In this note, we further examine the improvements proposed in [3], namely parallelization, adaptive proposals, and automatic partitioning of the space. The primary idea of simulated tempering is to sample from a tempered distribution $\pi_T(x) = \pi(x)^{1/T}$ for some temperature T. The algorithm proceeds by setting a temperature ladder $T = 1, \ldots, T_{\max}$ and running a Markov chain on the pair (x, T). As such, the chain explores the state space $\mathcal{X}$ while moving up and down the temperature ladder. Readers are referred to [4, 5] for further details. Of note for our purposes, however, is that one is able to specify pseudo-priors on the different steps of the ladder to ensure equal occupation numbers—time spent in each step of the ladder—which is a task well suited for stochastic approximation.

To test these (potential) improvements to simulated tempering, we employ a small bimodal density. Specifically, we set $\pi(x)$ to be an equally weighted mixture of two standard normal distributions, one centered at -15 and the other at 15. As such, the distribution has two modes (at $x = -15$ and $x = 15$) with a large low-density valley separating them. As a result, estimating the mean (0) is a natural

12 PAWL-Forced Simulated Tempering

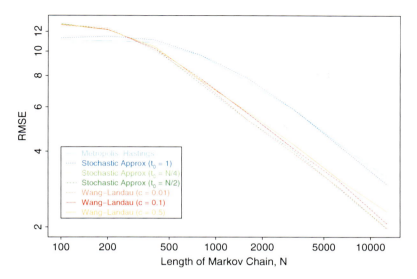

Fig. 12.1 RMSE for estimating the mean in the bimodal density for various simulated tempering configurations. We see that Wang–Landau (provided c is small) and stochastic approximation with deterministic step size decreases (provided t_0 is large) both perform well

challenge for any sampler. We run $1,000$ chains each of length N (for various N) and calculate the root mean squared error (RMSE) between the posterior mean (calculated from all states with $T = 1$) and the true mean of 0. We compare standard simulated tempering using Metropolis–Hastings with uniform pseudo-priors (using a Gaussian random walk with standard deviation 10, and temperatures $T = 1, 2, \ldots, 9, 10$) to that using stochastic approximation adjusted such that the pseudo-priors ensure equal occupation numbers. See [1] for details. We use standard stochastic approximation with step sizes $\gamma_t = t_0/\max(t_0, t)$ for $t_0 = 1, N/4, N/2$. In other words, the step size starts decreasing after 1 iteration, $N/4$ iterations, or $N/2$ iterations, respectively. We also explore Wang–Landau, which automatically decreases the step size after a flat histogram criterion is met. We look at three values of the user-specified tuning parameter c, namely $c = 0.01, 0.1, 0.5$. Figure 12.1 displays the RMSE as a function of N for each algorithm. We see that all of the stochastic approximation algorithms (including Wang–Landau) perform similarly in this simple example. It has been argued, however, that in more complex situations, Wang–Landau will outperform stochastic approximation with deterministically decreasing step size [1].

In Fig. 12.2 we similarly compare the simple Metropolis–Hastings simulated tempering algorithm to the Wang–Landau version (using $c = 0.1$) with and without adapting the proposal standard deviation (set to target an acceptance ratio of 0.234); see [3] for specifics. It is clear that adaptation in the proposal mechanism provides significant gains to both the standard simulated tempering algorithm as well as the Wang–Landau version. Further improvements might be made by considering

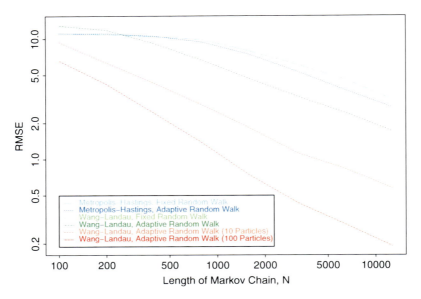

Fig. 12.2 RMSE for estimating the mean in the bimodal density for various simulated tempering configurations with and without adaptive proposals

mixture proposals tailored to each step on the temperature ladder, rather than being optimized to create a given acceptance rate across all temperatures. Figure 12.2 also displays the adaptive Wang–Landau algorithm in parallel with 10 and 100 particles, demonstrating vastly improved convergence of the algorithm. With M particles, the approximate improvement in RMSE is $\sqrt{M}$, which is roughly equivalent to if we were to run a single chain for $M \times N$ iterations. However, due to vectorization, the parallel version does not take M times as long to run. In our examples, $M = 10$ and $M = 100$ particles took 1.8 and 6.2 times longer than the single chain, respectively.

We also explored automatic setting of the temperature ladder using the bin-splitting method proposed in [3] (not shown). However, in this small example, the advanced binning method performed similarly to simply fixing the temperature ladder to the integers $1, \ldots, 10$. We suspect that in more complicated settings where the results are more sensitive to the temperature ladder the automatic binning approach will bring additional benefit.

12.3 Conclusion

This brief note has employed a simple bimodal example to demonstrate the benefits of embedding adaptive proposals, parallelization, and automatic bin splitting within the simulated tempering framework. Due to space limitations, many pertinent references and ideas have been excluded, though the interested reader might follow the citation trail to further explore these algorithms. If there is a single takeaway, it is that sometimes "stacking" multiple computational techniques can lead to significant improvements in performance. In this case, parallelization and adaptive proposals provide significant improvements to simulated tempering with the Wang–Landau algorithm; additionally, they are straightforward to implement through the R package PAWL, available online.

Ongoing work involves applying these simulated tempering methods to learn latent dimensions in nonstationary spatial models [2], which due to partial identifiability of the parameter space show particular promise for benefiting from the ideas presented herein. Specifically, as this class of models is new and as yet poorly understood, it is unclear a priori how to determine the scale of the proposal distribution as well as set the temperature ladder.

References

1. Atchade Y, Liu J (2010) The Wang-Landau algorithm for Monte Carlo computation in general state spaces. Stat Sin 20:209–233
2. Bornn L, Shaddick G, Zidek J (2012) Modeling nonstationary processes through dimension expansion. J Am Stat Assoc 107(497):281–289
3. Bornn L, Jacob PE, Del Moral P, Doucet A (2013) An adaptive interacting Wang-Landau algorithm for automatic density exploration. J Comput Graph Stat 22(3):749–773
4. Geyer C, Thompson E (1995) Annealing Markov chain Monte Carlo with applications to ancestral inference. J Am Stat Assoc 90(431):909–920
5. Marinari E, Parisi G (1992) Simulated tempering: a new Monte Carlo scheme. Europhys Lett 19(6):451
6. Wang F, Landau DP (2001) Efficient, multiple-range random walk algorithm to calculate the density of states. Phys Rev Lett 86(10):2050–2053

Chapter 13
Approximate Bayesian Computation for the Elimination of Nuisance Parameters

Clara Grazian

Abstract We propose a novel use of the approximate Bayesian methodology. ABC is a way to handle models for which the likelihood function may be considered intractable; this situation is closely related to the problem of the elimination of nuisance parameters: the model may contain a high-dimensional latent structure, so any elaboration of the likelihood function could be difficult or even impossible when the analysis is focused just on few parameters. We propose to use ABC to approximate the likelihood function of the parameter of interest.

13.1 Introduction

Recent developments allow Bayesian analysis also when the likelihood function $L(\theta; \mathbf{y})$ is intractable; that means it is analytically unavailable or computationally prohibitive to evaluate, for instance, because of a too high dimension of a latent structure that is part of the model. These methods are known as "approximate Bayesian computation" (ABC) or likelihood-free methods and are characterized by the fact that the approximation of the posterior distribution is obtained without explicitly evaluating the likelihood function. This kind of analysis was first proposed in genetics by [6], but it is now applied in a wide variety of settings, for instance, in finance, biology, physics, and signal processing.

The idea underlying likelihood-free methods is to propose a candidate θ' and to generate a data set from the working model with parameter set to θ'. If the observed and the simulated data are similar "in some way," then the proposed value is considered a good candidate to have generated the data and becomes part of the

C. Grazian (✉)
Dipartimento di Scienze Statistiche, University of Rome "La Sapienza", Rome, Italy
e-mail: clara.grazian@uniroma1.it

sample which will form the approximation to the posterior distribution. Conversely, if the observed and the simulated data are too different, the proposed θ' is discarded.

The basic version of the algorithm includes in the posterior sample all the proposal parameters that lead to a distance ρ between a suitable summary statistics $\eta(\cdot)$ computed on both the observed and the simulated data smaller than a tolerance level $\epsilon > 0$. If $\eta(\cdot)$ is a sufficient statistics, then

$$\lim_{\epsilon \to 0} \pi_\epsilon(\theta; \eta(\mathbf{y})) = \pi(\theta; \eta(\mathbf{y})). \tag{1}$$

The basic ABC may be inefficient; therefore, ABC algorithms are often linked with other methods, for instance, with Markov chain Monte Carlo (MCMC) or sequential Monte Carlo (SMC) methods. For a survey on the possible extensions to the original ABC algorithms and some applications, the reader may refer to [4, 5].

13.2 The Elimination of Nuisance Parameters

In a non-Bayesian setting, the problem of eliminating the nuisance parameters has no general solution. The idea underlying the available procedures (see [2]) is to accept a partial loss of data information: for instance, marginal or conditional experiments, in general, neglect a part of the likelihood which depends on the parameter of interest and the profile likelihood does not account uncertainty on the nuisance parameter; in general, they require the complete likelihood to be not too complex: partial experiments need to identify a function only dependent on the parameter of interest and the integrated likelihood [3] is based on an integral that may be neither analytically computed nor numerically approximated if the likelihood function is intractable.

Since the marginal posterior distribution for θ in the presence of a nuisance parameter ϕ is defined as

$$\pi(\theta; \mathbf{y}) = c\,\pi(\theta)\,L(\theta; \mathbf{y}) = c \int_\Phi \pi(\theta; \phi)\pi(\phi)\,L(\theta, \phi; \mathbf{y})d\phi \tag{2}$$

the likelihood function for the parameter θ may be rewritten as

$$L(\theta; \mathbf{y}) \propto \frac{\pi(\theta; \mathbf{y})}{\pi(\theta)} = \frac{\int_\Phi \pi(\theta, \varphi; \mathbf{y})\,d\varphi}{\int_\Phi \pi(\varphi)\,\pi(\theta; \varphi)\,d\varphi}. \tag{3}$$

Using ABC we can obtain an approximation of $\pi(\theta; \mathbf{y})$ constituted by a set of values which may be considered a sample from the (marginal) posterior distribution. In addition, provided the prior is proper, we are always able to simulate from it. With both a sample from the posterior distribution and a sample from the prior distribution, we can compute an approximation of the likelihood through the ratio of their density estimates.

13.2.1 Examples

In our work, we discuss some applications of the proposed method. In all of them, we have used simulated data and chosen as ABC parameter the Euclidean distance to compare sufficient statistics of the data and the data generated in the ABC step. We have always compared different tolerance levels.

First, we have analyzed the ABC approximation of the likelihood in situations where other solutions exist: one case where the parameter of interest is a transformation of the parameters of two independent Poisson distributions and one case from the class of Neyman and Scott, from [3]; the results are always good approximations of the integrated likelihood function. In general, the tolerance level seems to be a matter of computational power: when ϵ becomes smaller, the approximation is closer to the integrated likelihood; nevertheless smaller values are associated to higher computational costs.

Finally, we have used the ABC methodology to handle a class of problems with no straightforward solution, that is, the pseudo-likelihood for the quantiles of a distribution. In particular, we have analyzed one kind of quantile distribution: this is a class of distributions defined by their quantile functions that are nonlinear transformations of the quantiles of a standard normal distribution. For example, the quantile function of the g-and-k distribution presented in [1] is

$$Q_{gk} = (u;\ A,\ B,\ g,\ k) = A + B \left[1 + c\frac{1 - \exp\{-g\,z(u)\}}{1 + \exp\{-g\,z(u)\}}\right] \left\{1 + z(u)^2\right\}^k z(u), \tag{4}$$

where $z(u)$ is the uth standard normal quantile and A, B, g, and k represent location, scale, skewness, and kurtosis parameters, respectively.

This is a class of distributions characterized by a great flexibility of shapes obtained by varying parameters' values, which may model kurtotic or skewed data with the great advantage that they have a small number of parameters, unlike mixture models which are usually adopted to describe this kind of data. It is clear that the density function and then the likelihood function are unavailable; therefore any approach but ABC is difficult to implement. In this application, we have compared the basic ABC with ABC-MCMC; these algorithms have different computational costs, but it is evident from Fig. 13.1 that the algorithm with the MCMC step leads to a better approximation and is more efficient; however, they both lead to approximations concentrated around the true values of the considered quantiles. Again, the effect of choosing smaller tolerance levels is to obtain approximations closer to the true values of the quantiles of interest.

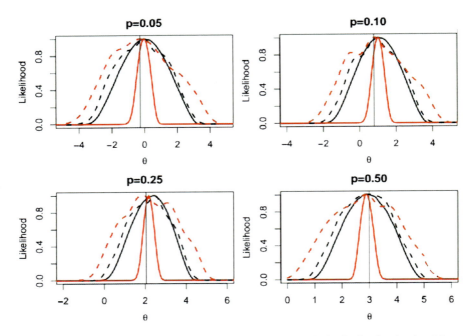

Fig. 13.1 Likelihood approximations of the quantiles of a g-and-k distribution for simulated data, obtained with both basic ABC (*dotted lines*) and ABC-MCMC with Gaussian transitional kernel (*solid lines*) algorithms and tolerance levels equal to 2 (*black lines*) and 0.5 (*red lines*). The parameter of interest θ is the quantile of level p

13.3 Conclusions

ABC is a very useful tool to handle a wide range of models, for which no other solution exists. In particular, it could be seen as a simple computational solution to the problem of eliminating nuisance parameters when the likelihood function is intractable and, therefore, classical solutions may not be used.

Nevertheless there remain many operational questions linked to the use of ABC algorithms regarding the choice of the summary statistics to use and how to perform the posterior approximations in the most efficient way. There are some recent suggestions to avoid the choices of ABC parameters: for instance, [7] proposes to use ABC linked with the empirical likelihood. ABC methods are currently under analysis to find both theoretical and practical advances.

A possible extension on which further research may be focused is an application of the proposed method in a nonparametric setting: in this case, ABC allows to manage highly flexible (and realistic) models with the only requirement to be able to simulate from them.

References

1. Allingham D, King RAR, Mengersen KL (2009) Bayesian estimation of quantile distributions. Stat Comput **19**(2):189–201
2. Basu D (1977) On the elimination of nuisance parameters. J Am Stat Assoc 72(358):355–366
3. Berger JO, Liseo B, Wolpert RL (1999) Integrated likelihood methods for eliminating nuisance parameters. Stat Sci 14(1):1–28
4. Blum M, Csilléry K, François O, Gaggiotti O (2010) Approximate Bayesian computation (ABC) in practice. Trends Ecol Evol 25(7):410–418
5. Marin JM, Pudlo P, Robert CP, Ryder R (1999) Approximate Bayesian computational methods. Stat Comput 21(2):289–291
6. Marjoram P, Molitor J, Plagnol V, Tavaré S (2003) Markov chain Monte Carlo without likelihoods. Proc Natl Acad Sci 100(26):15324–15328
7. Mengersen KL, Pudlo P, Robert CP (2012) Bayesian computation via empirical likelihood. Proc Natl Acad Sci USA 110(4):1321–1326

Chapter 14
Reweighting Schemes Based on Particle Methods

Reinaldo Marques and Geir Storvik

Abstract Sequential Monte Carlo methods are widely used to deal with the intractability of complex models including state space models. Their aim is to approximate the distribution of interest by a set of properly weighted samples. To control the weight degeneracy, the resample step has been proposed as an inexpensive alternative to avoid the collapse of particle filter algorithms. When the sample becomes too poor with successive use of resample steps, MCMC moves have been added in particle filter algorithms in order to make the identical samples diverge. In this work we consider strategies where we first perform a moves step, and then we update the weights for reweighting the particles. The validity of this approach is based on the commonly used trick of working on an artificial extended distribution having the target distribution as marginal combined with the use of backwards kernels. By updating the weights via a diversification step, this approach can make their empirical distribution less skewed increasing the effective sample size.

14.1 Introduction

State space models provide flexible representations for stochastic dynamical systems in which they can encapsulate many real problems (in time or space-time). This class of models is expressed in terms of an initial distribution, $\mathbf{x}_1 \sim \pi_\theta(\mathbf{x}_1)$, a state Markov model, $\mathbf{x}_t|\mathbf{x}_{1:t-1} \sim \pi_\theta(\mathbf{x}_t|\mathbf{x}_{t-1})$, and the observation model, $\mathbf{y}_t|\mathbf{y}_{1:t-1}, \mathbf{x}_{1:t} \sim \pi_\theta(\mathbf{y}_t|\mathbf{x}_t)$. All static parameters are represented by $\theta \in \Theta$, and for simplicity of notation we assume each $\mathbf{x}_t$ to have a common sample space $\mathscr{X} \subset \mathbf{R}^{p_x}$.

R. Marques (✉) • G. Storvik
University of Oslo and Statistics for Innovation Center, Oslo, Norway
e-mail: ramarque@math.uio.no; geirs@math.uio.no

Given the data $\mathbf{y}_{1:t}$ up to time t and assuming θ is known, inference focuses on the posterior distribution

$$\pi_t(\mathbf{x}_{1:t}) \equiv \pi_\theta(\mathbf{x}_{1:t}|\mathbf{y}_{1:t}) \propto \pi_\theta(\mathbf{x}_1)\pi_\theta(\mathbf{y}_1|\mathbf{x}_1) \prod_{k=2}^{t} \pi_\theta(\mathbf{x}_k|\mathbf{x}_{k-1})\pi_\theta(\mathbf{y}_k|\mathbf{x}_k). \quad (1)$$

When $\pi_t(\mathbf{x}_{1:t})$ is intractable, sequential Monte Carlo methods can be applied to carry out an approximate inference. Their respective algorithms—called *particle filters* (PF)—are exhaustively used to deal with the computational intractability of general state space models (see [1, 2, 7] and the references therein). The goal of such algorithms is to approximate the posterior by a set of properly weighted samples. The population of samples are called particles, and they are typically generated sequentially from some low-dimensional conditional distributions:

$$q_t(\mathbf{x}_{1:t}) = q(\mathbf{x}_1) \prod_{k=2}^{t} q(\mathbf{x}_k|\mathbf{x}_{k-1}). \quad (2)$$

Assuming that $q_t(\mathbf{x}_{1:t}) > 0$ for all $\mathbf{x}_{1:t}$ with $\pi_t(\mathbf{x}_{1:t}) > 0$, then based on sequential importance sampling ideas, particle weights are defined as

$$w_t(\mathbf{x}_{1:t}) \equiv \frac{\pi_t(\mathbf{x}_{1:t})}{q_t(\mathbf{x}_{1:t})} \propto w_{t-1} \frac{\pi_\theta(\mathbf{x}_t|\mathbf{x}_{t-1})\pi_\theta(\mathbf{y}_t|\mathbf{x}_t)}{q(\mathbf{x}_t|\mathbf{x}_{t-1})} \quad (3)$$

allowing for recursive computation. Estimation is usually based on normalized weights, making the proportionality constant unnecessary to compute.

A well-known problem in SMC methods is *weight degeneracy*, also called sample impoverishment. This problem is related to the increasing variance of the particle weights over time. In order to attenuate the degeneracy problem, [5] suggested to add resampling steps. Later, [4] proposed to add MCMC moves after the resampling step to reduce the sample impoverishment. This approach is called the resample-move (RM) algorithm, and the main idea is to create a greater diversity in the sample by rejuvenating the particles via a combination of sequential importance resampling and MCMC sampling steps. For this scheme, a Markov kernel $K_t(.|\mathbf{x}_{1:t})$ with $\pi_t(\mathbf{x}_{1:t})$ as the stationary distribution is designed to draw samples after the resample steps.

14.2 Particle Move-Reweighting Strategies

Assume $\mathbf{x}_{1:t} \sim q_t(\mathbf{x}_{1:t})$ is followed by a move $\mathbf{x}_{1:t}^\star \sim K_t(\mathbf{x}_{1:t}^\star|\mathbf{x}_{1:t})$ where K_t is π_t-invariant. As suggested by [3, 8] define an extended target distribution as $\bar{\pi}_t(\mathbf{x}_{1:t}^\star, \mathbf{x}_{1:t}) \equiv \pi_t(\mathbf{x}_{1:t}^\star) h_t(\mathbf{x}_{1:t}|\mathbf{x}_{1:t}^\star)$, where h_t is an artificial density/backward kernel that integrates to one. Following ordinary importance sample theory working on the enlarged space, we obtain the following result:

Proposition 1. Let $\{(\mathbf{x}_{1:t}^i, w_t(\mathbf{x}_{1:t}^i)), i = 1, \ldots, N\}$ be a properly weighted sample with respect to π_t where the particles are generated from q_t. Assume we make a move by a π_t-invariant MCMC kernel K_t to $\mathbf{x}_{1:t}^\star$ and update the weights by

$$\bar{w}_t(\mathbf{x}_{1:t}; \mathbf{x}_{1:t}^\star) = w_t(\mathbf{x}_{1:t}) \times r_t(\mathbf{x}_{1:t}; \mathbf{x}_{1:t}^\star),$$

where

$$r_t(\mathbf{x}_{1:t}; \mathbf{x}_{1:t}^\star) = \frac{\pi_t(\mathbf{x}_{1:t}^\star) h_t(\mathbf{x}_{1:t}|\mathbf{x}_{1:t}^\star)}{\pi_t(\mathbf{x}_{1:t}) K_t(\mathbf{x}_{1:t}^\star|\mathbf{x}_{1:t})},$$

and h_t is a density such that $\{(\mathbf{x}_{1:t}^\star, \mathbf{x}_{1:t}) : \pi_t(\mathbf{x}_{1:t}^\star) h_t(\mathbf{x}_{1:t}|\mathbf{x}_{1:t}^\star) > 0\}$ is a subset of $\{(\mathbf{x}_{1:t}^\star, \mathbf{x}_{1:t}) : q_t(\mathbf{x}_{1:t}^\star) K_t(\mathbf{x}_{1:t}|\mathbf{x}_{1:t}^\star) > 0\}$. Then

$$\{(\mathbf{x}_{1:t}^{\star i}, \bar{w}_t(\mathbf{x}_{1:t}^i; \mathbf{x}_{1:t}^{\star i})), i = 1, \ldots, N\}$$

becomes proper with respect to π_t.

For a proof, see [6].

By adding the MCMC move and updating the particle weights, we have that, for any density h, $\bar{w}_t$ is a proper weight for $\mathbf{x}_{1:t}^\star$ with respect to π_t. Even though we are working with an extended space, the MCMC move does not modify the unbiased property. The main point of this sampling scheme is to highlight the multiple choices of proper weight functions $\bar{w}_t(\mathbf{x}_{1:t}; \mathbf{x}_{1:t}^\star)$ for any given kernel K_t. For specific choice of h_t, $\bar{w}_t$ reduces to w_t. Also, for clever choices of the artificial density, we can obtain that $Var[\bar{w}_t] < Var[w_t]$ allowing us to reduce the weight degeneracy and simultaneously to create sample diversity. Some special choices of the artificial density and moves from arbitrary transitions kernels are discussed in [6].

When we have a properly weighted sample, there are some interesting SMC schemes to design, taking into different types of moves. In short, we can adduce at least four strategies for (re)weighting the particles in an SMC framework:

s.1 Start with an equally weighted sample, move the particles via an MCMC kernel, and keep the equal particle weights (the *resample-move* approach).
s.2 Start with a properly weighted sample, move the particles via an MCMC kernel, and keep the same particle weights.
s.3 Start with a properly weighted sample, move the particles via an MCMC kernel, and update the particle weights.
s.4 Start with a properly weighted sample, move the particles via an arbitrary kernel, and update the particle weights.

The Table 14.1 summarizes results for the following spate space model: $y_t \sim Poisson(e^{\mathbf{b}\mathbf{x}_t + \lambda})$ and $\mathbf{x}_t \sim \mathcal{N}(\mathbf{F}\mathbf{x}_{t-1}, \Sigma)$. All static parameter we assume to be known. The table shows that updating the weights via a MCMC move, we can obtain a considerable increase in the effective sample size and an improvement in the predictive performance.

Table 14.1 Results of resample-move (RM) and move-reweighting (MR) algorithms

Algorithms	ESS (%)	RMSE
RM	26	94.54
MR	67	84.40
MR-RM	70	83.38

The table shows the average over time of the effective sample size (ESS) and root mean square root (RMSE) for 50 particles and sample size equal to 200

14.3 Closing Remarks

In this work, we propose a flexible strategy that allows for MCMC moves without the need of a preliminary resampling step. Following the MCMC moves, we update the particle weight taking into account the diversification step and that MCMC moves give particles closer to the target distribution. Our approach allows a great flexibility to reweight the particles using a proper weight with respect to the right distribution. The main effect of this is that we can increase the effective sample size; consequently the resample stages can be delayed.

Acknowledgements We gratefully acknowledge financial support from CAPES-Brazil and Statistics for Innovation Center in Norway.

References

1. Andrieu C, Doucet A, Holenstein R (2010) Particle Markov chain Monte Carlo methods. J Roy Stat Soc Ser B (Stat Methodol) 72(3):269–342
2. Chopin N, Jacob P, Papaspiliopoulos O (2012) SMC^2: an efficient algorithm for sequential analysis of state space models. J Roy Stat Soc Ser B (Stat Methodol). doi:10.1111/j.1467-9868.2012.01046.x
3. Del Moral P, Doucet A, Jasra A (2006) Sequential Monte Carlo samplers. J Roy Stat Soc Ser B (Stat Methodol) 68(3):411–436
4. Gilks W, Berzuini C (2001) Following a moving target: Monte Carlo Inference for dynamic Bayesian models. J Roy Stat Soc Ser B (Stat Methodol) 63(1):127–146
5. Gordon N, Salmond D (1993) Novel approach to nonlinear/non-Gaussian Bayesian state estimation. Radar Signal Process, IEE Proc F 140(2):107–113
6. Marques R, Storvik G Online inference for dynamical models: the move-reweighting particle filter (in preparation)
7. Robert C, Casella G (2004) Monte Carlo statistical methods. Springer, New York
8. Storvik G (2011) On the flexibility of metropolis–hastings acceptance probabilities in auxiliary variable proposal generation. Scand J Stat 38(2):342–358

Chapter 15
A Bayesian Nonparametric Framework to Inference on Totals of Finite Populations

Juan Carlos Martínez-Ovando, Sergio I. Olivares-Guzmán, and Adriana Roldán-Rodríguez

Abstract In this chapter we sketch a Bayesian model-based framework to inference on totals of finite populations. Our formulation takes into account a population that is partitioned into planned domains or strata. The inferential framework is based on the decomposition of the population total into sampled and unsampled parts. Inference on the unsampled part of the total is made using Bayesian nonparametric methods, in the absence of design information on unsampled individuals.

15.1 Introduction

In this chapter we sketch a novel model-based framework to addressing an inferential problem frequently appeared in finite population studies. That is to make inference on totals of a finite population. In our formulation, it is assumed that the population of interest is being divided into planned domains or strata. It is also assumed that the characteristic to be measured in each individual is random and continuous.

The population of interest is being denoted by $\mathscr{P}$. It is assumed that $\mathscr{P}$ is partitioned into J planned domains, $\{\mathscr{P}_j\}_{j=1}^{J}$. It is also assumed that the number of individuals belonging to each planned domain is known. Accordingly, the total of $\mathscr{P}$ can be decomposed as the sum of J independent partial totals,

$$T = \sum_j T_t, \qquad (1)$$

J.C. Martínez-Ovando (✉) • S.I. Olivares-Guzmán • A. Roldán-Rodríguez
Dirección General de Investigación Económica, Banco de México, México D. F., México
e-mail: juan.martinez@banxico.org.mx; solivares@banxico.org.mx; aroldan@banxico.org.mx

where $T_j = \sum_{l=1}^{N_j} Y_{jl}$, and Y_{jl} is the characteristic of the lth individual. Here, N_j stands for the number of individuals in $\mathscr{P}_j$. Given the above decomposition, inferences on T are made aggregating inferences of each partial total T_j.

15.2 Inference on Planned Domains

Our formulation makes use of an intuitive decomposition of each T_j into sampled and unsampled parts, as

$$T_j = T_j^{\mathscr{S}} + T_j^{\tilde{\mathscr{S}}}, \qquad (2)$$

where $\mathscr{S}_j$ and $\tilde{\mathscr{S}}_j$ stand for the sampled and unsampled parts of $\mathscr{P}_j$, respectively. Once the sampled part $\mathscr{S}_j$ is being observed, inference on the unsampled part is made using Bayesian nonparametric methods, see [1, 3, 8] for previous formulations of the problem based on Dirichlet processes and mixtures of Dirichlet processes, respectively. They implicitly assume no design information on unsampled individuals. In our formulation, we preserve that assumption under the auspice of the notion of exchangeability within planned domains (see, e.g. [5]).

The Bayesian nonparametric component we choose for our formulation belongs to the class of *species-sampling models* (SSMs); see [6]. Under the marginalization property, SSMs express the individual predictive distribution as a weighted average of the sampled data and a baseline distribution function. See [2]. The predictive distribution induced by SSMs is thus interpretable and analytically tractable.

15.2.1 Posterior Point Estimates

Under SSMs, predictive point estimates of totals on planned domains have a simple and intuitive expression, as

$$\hat{T}_j = \sum_{l \in \mathscr{S}_j} y_{jl} + N_j^{\tilde{\mathscr{S}}} \cdot \left[\sum_{k=1}^{U_j} \left(\rho_k(\boldsymbol{m}_j) \cdot y_{jk}^* \right) + \phi(\boldsymbol{m}_j) \cdot \widehat{\mu_{j0}} \right], \qquad (3)$$

where U_j is the number of unique measurements (or ties) in $\mathscr{S}_j$, $\boldsymbol{y}_j^* = \{y_{jk}^* : k = 1, \ldots, U_j\}$ is the collection of those sampled ties, $N_j^{\tilde{\mathscr{S}}}$ is the number of unsampled individuals in $\mathscr{P}_j$, and $\widehat{\mu_{j0}} = \mathrm{E}_{G_{j0}}\{Y_{jl}|\boldsymbol{\theta}_{j0}\}$ is the prior expectation of an individual measurement Y_{jl} in $\tilde{\mathscr{S}}_j$. This expectation is computed with respect to the baseline distribution G_{j0}, with $\boldsymbol{\theta}_{j0}$ being its indexing parameter.

In the above expression, $\boldsymbol{m}_j$ denotes the vector of sampled frequencies attained to the collection of ties, $\boldsymbol{y}_j^*$. Additionally, (ρ_k) and ϕ are positive functions, such that $\sum_{k=1}^{U_j} \rho_k(\boldsymbol{m}_j) + \phi(\boldsymbol{m}_j) = 1$.

As [1] noticed before, (3) encompasses traditional weight-based stratified and post-stratified point estimates (see [7]). However, our formulation allows us to produce full posterior distribution of each T_j and relevant aggregations of them.

15.2.2 Full Posterior Inference

Full posterior inferences on T_j are obtained in terms of the distribution of the unsampled part of the total, $T_j^{\widetilde{\mathscr{S}}}$. The predictive distribution for each T_j is obtained as a shifted $N_j^{\widetilde{\mathscr{S}}}$-fold convolution distribution,

$$\Pr(T_j \leq t | \mathscr{S}_j) = \widehat{G}_j^{*N_j^{\widetilde{\mathscr{S}}}} \left(T_j^{\widetilde{\mathscr{S}}} \leq t + T_j^{\mathscr{S}} \right), \qquad (4)$$

with shifting constant $T_j^{\mathscr{S}} = \sum_{l \in \mathscr{S}_j} y_{jl}$. Notice that the above convolution is computed in terms of $\widehat{G}_j$, the individual predictive distribution induced by the SSM.

Although the analytic expression for (4) may be intricate, it is possible to reproduce it using stochastic simulation techniques. It is worth to mention, as well, that inference on the total of any aggregation of planned domains can be easily produced through simulation.

15.3 Simulation Results

Figure 15.1 displays the results of a simulation study consisting of augmenting sampling schemes of a simulated population with two planned domains. Actual totals are indicated with the red-coloured vertical lines. Uncertainty surrounding the planned domain totals dissipates as the sample size increases.

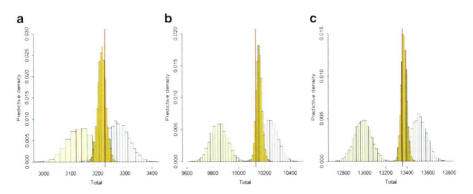

Fig. 15.1 Prediction on disaggregated and population totals for samples of variable sizes. Panels: (**a**) planned domain 1, (**b**) planned domain 2 and (**c**) whole population

15.4 Discussion

In this chapter we have sketched a Bayesian nonparametric framework to make inference on totals of finite population, based on individual continuous measurements. A key distinction of our framework with regards to traditional design-based alternatives is that the characteristic of interest of each individual is assumed as random. Thus full and interpretable posterior inferences on totals using convolution-type distributions are doable. The results presented here are part of a broader research agenda to make Bayesian nonparametric inference on finite populations, assuming non-informative sampling schemes. Further and more detailed results are reported in [4].

Acknowledgements The first author gratefully acknowledges an SNI research stimulus from CONACYT (Mexico). The views expressed in this article are those of the authors and do not necessarily reflect those of Banco de México.

References

1. Binder DA (1982) Non-parametric Bayesian models for samples from finite populations. J Roy Stat Soc Ser B 44:388–393
2. Hansen B, Pitman J (2000) Prediction rules for exchangeable sequences related to species sampling. Stat Probab Lett 46(3):251–256
3. Lijoi A, Prünster I (2010) Models beyond the Dirichlet process. In: Bayesian nonparametrics. Cambridge University Press, Cambridge, pp 80–130
4. Martínez-Ovando JC, Olivares-Guzmán SI, Roldán-Rodríguez A (2013) Predictive inference on finite populations based on planned and unplanned domains. Mimeo, Discussion paper, Banco de México
5. Meeden G, Vardeman S (1991) A noninformative Bayesian approach to interval estimation in finite population sampling. J Am Stat Assoc 86:972–980
6. Pitman J (1995) Exchangeable and partially exchangeable random partitions. Probab Theory Relat Fields 102(2):145–158
7. Thompson ME (1997) Theory of sample surveys. Monographs on statistics and applied probability. Chapman & Hall, London
8. Zangeneh SZ, Keener RW, Little R (2011) Bayesian nonparametric estimation of finite population quantities in absence of design information on nonsampled units. In: Proceedings of the joint statistical meeting – section on survey research methods. American Statistical Association, Alexandria, pp 3429–3437

Chapter 16
Parallel Slice Sampling

Teresa Pietrabissa and Simone Rusconi

Abstract To draw a sample of a continuous variable B from a finite measure g, using a Markov chain Monte Carlo (MCMC) method, there is an easy algorithm named slice sampling. The two main problems of this algorithm are the solution of a inequality, involving the measure density g, which can be hard to find due to the irregularities g can be affected by, and the high dimensionality of the support of the density itself.

Our aim is to create a library, using the slice sampling, to draw an MCMC sample from any density g, in particular to get a realization from the posterior density of a Bayesian model. We will present and discuss a solution and some statistical test applications, using a GPU parallel language, which is, nowadays, becoming more and more commonly employed in the context of Bayesian statistical models.

16.1 Introduction

MCMC methods provide a general approach for approximating integrals with respect to a wide range of complex distributions.

In a typical Bayesian analysis, the integration measure is the posterior distribution. Let $\underline{\beta} \in \mathbb{R}^k$ be the argument of the posterior, i.e., the variable of integration. The posterior distribution usually has a density with respect to the Lebesgue measure (for instance) and is proportional to a measure density:

$$g(\underline{\beta}) = L(\underline{\beta}|X)\pi(\underline{\beta}), \qquad (1)$$

where $L(\underline{\beta}|X)$ is the likelihood, seen as function of $\underline{\beta}$ and evaluated in the data X, and $\pi(\underline{\beta})$ is the prior density of the model. Our purpose is to create a library

---

T. Pietrabissa (✉) • S. Rusconi
Department of Mathematics, Politecnico di Milano, Milan, Italy
e-mail: teresa.pietrabissa@mail.polimi.it; simone1.rusconi@mail.polimi.it

that takes as input a generic $g(\underline{\beta})$ and returns an MCMC sample from the posterior density:

$$\underline{\beta}^{(i)} \text{ with } i = 1 \div G. \tag{2}$$

Finally, as applications, we will compute posterior distributions in some "benchmark" Bayesian models using parallel slice sampling. In particular we will test the goodness of our algorithm on some conjugate models comparing theoretical posterior distribution density with the resulting sample; moreover, we will use it to study the well-known dataset "quakes" available in the R software.

16.2 The Algorithm

The algorithm that we are going to follow is the slice sampling. This is a simple Gibbs-Sampler, which adds to the variables space the continuous scalar random variable U. The Gibbs-Sampler full conditionals are

$$(U|B = \beta) \sim \mathscr{U}(0; g(\beta)) \tag{3}$$

$$(B|U = u) \sim \mathscr{U}(S) \text{ where } S = \{\beta : g(\beta) > u\}. \tag{4}$$

So the algorithm can be summarized as follows:
  After initializing $\underline{\beta}^{(0)}$ as a suitable value:

1. Sample $u^{(i)}$ from $\mathscr{U}(0; g(\underline{\beta}^{(i)}))$.
2. Sample $\underline{\beta}^{(i+1)}$ from $\mathscr{U}(S^{(i)})$, where $S^{(i)} := \{\underline{\beta} \in \mathbb{R}^k | g(\underline{\beta}) > u^{(i)}\}$.

  Repeat steps 1 and 2 for $i = 0 \div G - 1$ to obtain the sample desired.

MCMC methods are sequential methods as their structure requires it. The same is for the slice sampling where, as briefly shown, we get a new realization from the previous one. Indeed, we are sampling a Markov chain, which is time dependent by definition.

However, each evaluation of the new realization $\underline{\beta}^{(i+1)}$ can be split into simple steps. Step 1 is easily computable in a sequential way on CPU, but step 2 requires the grate computing power of GPU architectures.

More details on the two main steps are given in the following subsections.

16.2.1 First Step

The first step requires to sample from $\mathscr{U}(0; g(\underline{\beta}^{(i)}))$, with $\underline{\beta}^{(i)}$ known. This is the easiest step of the algorithm as it consists in sampling from a probability distribution which is simple to generate from. Moreover, this is a one-dimensional random number generation and, for this reason, it becomes easier to run it with libraries that fully support one-dimensional random number generation itself.

16.2.2 Second Step

The second step consists in sampling $\beta^{(i+1)}$ from $\mathscr{U}(S^{(i)})$, which requires resolution of inequality $g(\beta) > u^{(i)}$. Based on the computing power of GPU programming, we follow these simple steps:

1. In a parallel way, sample $\underline{\beta}_{(j)}^{(i)}$ from $\mathscr{U}(R)$, where $j = 1 \div K$, K is as big as needed, i is the fixed main step index, $R \subset \mathbb{D}$ is a k-dimensional hyper-rectangle, and $\mathbb{D} \subseteq \mathbb{R}^k$ is the domain of the function $g(\beta)$.
2. In a parallel way, check if some $\underline{\beta}_{(j)}^{(i)}$ make the value of $g(\underline{\beta}_{(j)}^{(i)})$ overcome the value of $u^{(i)}$.
3. Repeat steps 1 and 2 until at least one of the $\underline{\beta}_{(j)}^{(i)}$ verifies the condition required at previous step.
4. The new realization $\underline{\beta}^{(i+1)}$ is equal to one of the $\underline{\beta}_{(j)}^{(i)}$ found at step 2.

The hyper-rectangle R can be obtained in different ways and in general we will ask the user to provide it. Once we have defined R, we can sample from k one-dimensional uniforms, one for each dimension. This way we are able to sample $\underline{\beta}_{(j)}^{(i)}$ from $\mathscr{U}(R)$.

Summarizing, we try to solve the inequality $g(\beta) > u$ with a brute-force attack. In this approach are needed two contrasting features: K, the number of attempts, must be very large, but the computational time should remain low. A way to reconcile these two aspects is to take advantage of the great computing power of GPU architectures.

16.3 A Simple Example

We present here a simple example on how to use our program and what is the output obtained. To do so we tried to get an MCMC sample from a beta-binomial distribution, as we already know that the prior density

$$\pi(\beta) = \text{Beta}(a, b)$$

is conjugate with respect to the binomial likelihood, generating the posterior density:

$$pi(\beta \mid \underline{X}) = \text{Beta}\left(a + \sum_{i=0}^{n} X_i, b + n - \sum_{i=0}^{n} X_i\right)$$

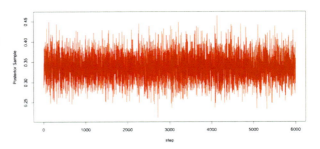

Fig. 16.1 Traceplot for the posterior sample, obtained from the beta-binomial model with burning = 50, thinning = 3

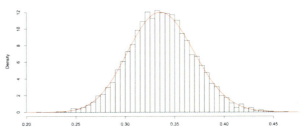

Fig. 16.2 Comparison in between theoretical distribution (*in red*) and the histogram of the posterior sample

Using R software we drew a sample from the likelihood:

$$\underline{X} \mid \beta \sim Bi(n, \beta) \text{ where } n = 200 \text{ and } \beta = 0.3$$

The prior density for the model, as already introduced, is the beta distribution

$$\beta \sim \pi(\beta) = \text{Beta}(a, b) \text{ where } a = b = 1.2$$

so that our prior is uninformative with respect to the parameter β.

Using our library, we got the posterior sample of the model and obtained what was expected from theoretical analysis.

As shown in the traceplot, the posterior sample has a very good behavior and it seems not to have any particular shape. If we compare the theoretical distribution with the histogram of the posterior sample, we find a very good match which states the goodness of the algorithm (Figs. 16.1 and 16.2).

Chapter 17
Approximate Bayesian Computation in Quantile Regression

Antonio Pulcini

Abstract We propose an approximate Bayesian approach to estimate the joint distribution of the response variable Y and the set of covariates $\mathbf{X}$ based on the notion of quantile distribution. We focus on cases where the quantile regression framework is necessary, but the unknown form of the regression function and the large number of quantiles suggest to directly estimate the conditional distribution. In these cases, the use of very flexibly shaped distributions may be of interest. In this context, we adopt the multivariate **g**-and-**h** distribution, a member of the quantile distribution family. Due to the lack of the likelihood function in a closed and manageable form, the estimation proceeds via an approximate Bayesian computation (ABC) algorithm that allows us to easily estimate all the parameters. The performance of the proposed approach is evaluated via simulated data sets.

17.1 Summary

The usual assumptions of the standard linear model imply that the conditional distribution of the response variable is, at least approximately, Gaussian. In practice, this assumption is rarely acceptable. In many observational studies, the conditional distribution is not symmetric and, even worse, its shape depends on the value of the covariates [7]. For these situations, methods based on quantile regression are common alternatives [4]. Nevertheless when influence concerns more than one quantile, it may be convenient to consider the problem of directly estimating the conditional distribution of the response variable given the explanatory variables [5]. In this context quantile distributions, due to their flexibility and the small number

A. Pulcini (✉)
Università di Roma TOR VERGATA, Rome, Italy
e-mail: antoniopulcini@gmail.com

of parameters, may represent a valid choice. Field and Genton [3] have proposed a generalization of the univariate **g**-and-**h** distribution to the multivariate case.

We exploit the use of the multivariate **g**-and-**h** distribution for the estimation of the joint distribution of the response variable Y given a vector of covariates $\mathbf{X} = (X_1, \ldots, X_K)$.

A drawback of quantile distributions, which has represented an obstacle to their use, is the lack of a closed-form expression of the likelihood function. On the other hand, the problem of generating random values from them is an easy task. These issues suggest the estimation via the approximate Bayesian computation (ABC) approach [1,6].

ABC allows to produce a sample from an approximate version of the posterior distribution. No likelihood evaluation is required, only a way to sample from the model distribution.

Briefly, we assume data, $\mathscr{D}$, arise from the multivariate **g**-and-**h** distribution:

$$W = \Sigma^{1/2} R_{g,h}(Z) + \mu, \qquad (1)$$

where:

- $\mu \in R^{K+1}$ is the location,
- Σ is the variance covariance matrix,
- $g = (g_1, g_2, \ldots, g_{K+1}) \in R^{K+1}$ controls the skewness,
- $h = (h_1, h_2, \ldots, h_{K+1}) \in R_+^{K+1}$ controls the kurtosis,
- $Z \sim N_{K+1}(0, I)$, and
- $R_{g,h}(Z) = (R_{g_1,h_1}(Z_1), R_{g_2,h_2}(Z_2), \ldots, R_{g_{K+1},h_{K+1}}(Z_{K+1}))^T$,

with $R_{g,h}(z) = \left(\frac{\exp(gz)-1}{g}\right) \exp\left(\frac{hz^2}{2}\right)$.

In order to estimate all the parameters in the model we propose the following ABC-MCMC algorithm:

1. Being at θ_t, propose a move to θ' according to a normal transition kernel $q(\theta_t \to \theta')$.
2. Generate M samples, $\mathscr{D}'_1, \ldots, \mathscr{D}'_M$, from the model with parameters θ'.
3. Calculate $\alpha = \min\left(1, \frac{\frac{1}{M}\sum_{m=1}^{M} K_\epsilon(\rho(S(\mathscr{D}),S(\mathscr{D}'_m)))\pi(\theta')q(\theta' \to \theta_t)}{\frac{1}{M}\sum_{m=1}^{M} K_\epsilon(\rho(S(\mathscr{D}),S(\mathscr{D}'_{m;t})))\pi(\theta_t)q(\theta_t \to \theta')}\right)$.
4. Accept θ' with probability α; otherwise stay at θ_t.
5. Update the variance of $q(\cdot)$ through an Adaptive MCMC scheme, then return to 1.

Specifically, $S(\cdot)$ is a multivariate quantile [2], $\rho(\cdot)$ the Euclidean norm, and $K_\epsilon(\cdot)$ a multivariate Gaussian kernel centered on $S(\mathscr{D}') = S(\mathscr{D})$ with variances ϵ's.

The performance of the proposed method is evaluated with simulated datasets. Secondly we apply our approach to estimate the joint distribution of price and demand in the Italian day-ahead electricity market.

References

1. Allingham D, King RAR, Mengersen KL (2009) Bayesian estimation of quantile distributions. Stat Comput 19:189–201
2. Chaudhuri P (1996) On a geometric notion of quantiles for multivariate data. J Am Stat Assoc 91:862–872
3. Field C, Genton MG (2006) The multivariate g-and-h distribution. Technometrics 48:104–111
4. Koenker R (2005) Quantile regression. Cambridge University Press, Cambridge
5. Peracchi F (2002) On estimating conditional quantiles and distribution functions. Comput Stat Data Anal 38:433–447
6. Tavaré S, Balding D, Griffiths RC, Donnelly P (1997) Inferring coalescence times from DNA sequence data. Genetics 145:505–518
7. Yu K, Lu Z, Stander J (2003) Quantile regression: applications and current research areas. The Statistician 52:331–350

Part III
Bayes @ Work: Appraisal of Applications to the Real World

Chapter 18
Spatiotemporal Model for Short-Term Predictions of Air Pollution Data

Francesca Bruno and Lucia Paci

Abstract Recently, the interest of many environmental agencies is on short-term air pollution predictions referred at high spatial resolution. This permits citizens and public health decision-makers to be informed with visual and easy access to air-quality assessment. We propose a hierarchical spatiotemporal model to enable use of different sources of information to provide short-term air pollution forecasting. In particular, we combine monitoring data and numerical model output in order to obtain short-term ozone forecasts over the Emilia Romagna region where the orography plays an important role on the air pollution; thus, the elevation is also included in the model. We provide high-resolution spatial forecast maps and uncertainty associated with these predictions. The assessment of the predictive performance of the model is based upon a site-one-out cross-validation experiment.

18.1 Introduction

Recently, several environmental agencies are interested to provide the public and experts with visual and easy access to air-quality information. Short-term air pollution predictions are usually needed as high-resolution spatial patterns. Numerical models devoted to estimate air pollution or meteorological variables are usually available in short-time frames at grid-cell spatial resolution. However, these forecasts are often biased and not equipped by any uncertainty measure. Conversely, measurements at monitoring stations give the "true" air pollution level. The joint consideration of the two sources of information can improve air pollution forecasting [5, 6].

F. Bruno • L. Paci (✉)
Department of Statistical Sciences "Paolo Fortunati", University of Bologna,
Via delle Belle Arti, 41 - Bologna, Italy
e-mail: francesca.bruno@unibo.it; lucia.paci2@unibo.it

In this work, ozone data from the monitoring network and numerical model are fused to obtain accurate short-term predictions of ozone level in the Emilia Romagna region, in Italy. We employ the downscaling approach described in [1] where point and grid-referenced data are combined via a linear regression model with spatiotemporally varying coefficients [4].

The monitoring data we use are collected on hourly basis at $n = 41$ stations during August, 2012. Moreover, two numerical models are available: the first one is the multi-scale Chimere chemistry-transport model which estimates the ozone level (and other pollutants) for gridded cells (at 5 km resolution) over successive time periods up to 72 h in the future. Also, the weather numerical model Cosmo is available. This is a non-hydrostatic atmospheric model developed by the Consortium for small-scale modeling. Cosmo model is run every day producing 72 h forecasts at 7 km spatial resolution for several meteorological variables. Here, we consider the hourly temperature forecasts produced by Cosmo at the grid cells spanning the region, since the temperature influences directly the kinetics of reactions producing ozone [2]. Finally, the orography of the region will be taken into account since the ozone level changes according to the elevation. This information is available at both point-level and grid-cell spatial resolution.

18.2 Data-Fusion Model

Let $Y_t(\mathbf{s})$ denote the hourly ozone concentration at a location $\mathbf{s}$, $A(\mathbf{s})$ be the altitude of site $\mathbf{s}$, and $C_t(B)$ define the numerical model output over the grid cell B (i.e., Chimere or Cosmo output). Following [1], we address the spatial misalignment between monitoring data and numerical model output, by associating to each site $\mathbf{s}$ the grid cell $A(\mathbf{s})$ that contains $\mathbf{s}$. Then, the model links the observational data and the numerical model output as follows:

$$Y_t(\mathbf{s}) = \beta_0 + \beta_{0,t}(\mathbf{s}) + \big(\beta_1 + \beta_{1,t}(\mathbf{s})\big) C_t(B) + \beta_2 A(\mathbf{s}) + \epsilon_t(\mathbf{s}), \tag{1}$$

where $\epsilon_t(\mathbf{s})$ is a white noise pure error process with nugget variance τ^2. The spatiotemporally varying coefficients $\beta_{0,t}(\mathbf{s})$ and $\beta_{1,t}(\mathbf{s})$ have a multiplicative form in temporal and spatial effects leading to independent zero-mean processes with separable covariance functions. However, this will not imply space-time separability of dependence structure for the $\mathbf{Y}$'s [3].

Predictions at new site $\mathbf{s}'$ and future hour t' are based upon the predictive distribution of $Y_{t'}(\mathbf{s}')$. Model (1) is fitted using a Gibbs sampler, using both the air-quality model output (Chimere) and the weather model output (Cosmo).

18.3 Analyses and Results

We illustrate by modeling data of 48-h running windows starting at each hour from 10 AM on August 10th to 9 AM on August 11th, 2012. A site-one-out

Table 18.1 MSPE for 3-h ahead ozone forecasts obtained from model (1) with Chimere, with Cosmo, and with Cosmo excluding elevation

	1-h	2-h	3-h
Chimere	364.81	568.92	825.93
Cosmo	278.97	350.47	461.14
Cosmo without $A(s)$	394.26	479.18	624.99

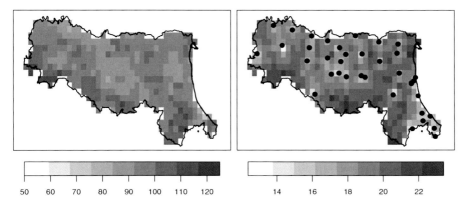

Fig. 18.1 The 1-h ahead ozone forecast map (*left panel*) and the standard deviation map (*right panel*) at 12 PM on 12th August in µg/m^3. *Black dots* represent monitoring sites

cross-validation experiment is performed to produce 1-h, 2-h, and 3-h ahead forecasts of ozone level for 24 consecutive windows. For example, we model data from 10 AM on August 10th to 9 AM on August 12 using $n - 1$ monitoring sites. Forecasts of ozone level are obtained at the validation site for 10 AM (1-h ahead), 11 AM (2-h ahead), and 12 PM (3-h ahead) on August 12th. This is repeated such that each monitoring station is used once as validation site.

Table 18.1 provides a comparison of out-of-sample 1-h, 2-h, and 3-h ahead ozone predictions. In terms of mean squared prediction error (MSPE), the model with Cosmo output as covariate outperforms the one fitted using the Chimere output. This is due to the high degree of smoothness of the air-quality model output. The inclusion of the altitude, $A(s)$, in model (1) improves the ozone forecasting. We also note that the MSPE tends to increase as the length of the forecast period increases, as we expected. Other indexes such as the normalized mean bias factor, the weighted normalized mean square error, and the correlation coefficient lead to similar results.

Finally, Fig. 18.1 (left panel) shows the 1-h ahead forecast map at 12 PM on August 12th, obtained as posterior predictive mean under model (1) fitted using the Cosmo output. The posterior standard deviation map in the right panel gives a measure of the uncertainty associated with the predictions.

Model (1) is simple, very flexible, and computationally efficient. Accurate ozone forecasts are obtained in short-time frames along with associated uncertainty. Therefore, it represents an attractive strategy for environmental agencies which usually do not adopt stochastic approaches.

Acknowledgements The research work underlying this paper was funded by a FIRB 2012 grant (project no. RBFR12URQJ) for research projects of national interest that was provided by the Italian Ministry of Education, Universities and Research. We also would like to thank ARPA-SIMC Emilia Romagna, for providing monitoring data set and the output of the numerical models Chimere and Cosmo.

References

1. Berrocal VJ, Gelfand AE, Holland DM (2010) A spatio-temporal downscaler for output from numerical models. J Agric Biol Environ Stat 14:176–197
2. Cocchi D, Trivisano C (2013) Ozone. In: El-Shaarawi AH, Piegorsch W (eds) Encyclopedia of environmetrics. Wiley, Chichester. doi: 10.1002/9780470057339.vao022.pub2
3. De Cesare L, Myers DE, Posa D (2001) Estimating and modeling space-time correlation structures. Stat Probab Lett 51:9–14
4. Gelfand AE, Kim H-J, Sirmans CF, Banerjee S (2003) Spatial modeling with spatially varying coefficient processes. J Am Stat Assoc 98:387–396
5. Paci L, Gelfand AE, Holland DM (2013) Spatio-temporal modeling for real-time ozone forecasting. Spat Stat 4:79–93
6. Sahu SK, Yip S, Holland DM (2009) A fast bayesian method for updating and forecasting hourly ozone levels. Environ Ecol Stat 18:185–207

Chapter 19
Predicting Rainfall Fields from Lightning Records: A Hierarchical Bayesian Approach

Edmondo Di Giuseppe, Giovanna Jona Lasinio, Massimiliano Pasqui, and Stanislao Esposito

Abstract Mixed models (linear and nonlinear) belong to a class of models in which some of the effects are *fixed* and some are *random*; formalization of these models is easily achieved in a hierarchical Bayesian framework. Here we propose a space-time mixed model to link rain measures and lightning counts in a given area of Central Italy.

19.1 Introduction

In this paper we aim at formulating a model for predicting the 15-min cumulated precipitation at unknown locations and time given lightning counts. In particular, we assume that the cumulated precipitation at time t in cell p of a 10×10 km regular grid is generated by a fixed component related to lightnings and a random **W** term structured in space and time. We refer to events during *convective storms*.

Table 19.1 Frequency distribution of observed values and discretization of the latent rainfall field

Rain classes (mm)	[0, 0.2)	[0.2, 0.4)	[0.4, 0.6)	[0.6, 0.8)	[0.8, 1)	≥ 1
Cases	9,328	861	353	258	199	1,173
Discretization values	$\lambda_0 = \log(0.1)$	$\lambda_1 = \log(0.3)$	$\lambda_2 = \log(0.5)$	$\lambda_3 = \log(0.7)$	$\lambda_4 = \log(0.9)$	

We use lightning records in the fixed component of the model. The study area is located in Central Italy; we analyze an event of 68 time units (May 9, 2006). The database is composed of lightning records (instant-point fields) cumulated over a grid with 10×10 km cells and 179 rain gages. When two or more rain gages belong to the same grid cell we take their median over the cell ending with 111 spatial measurements at each time point.

Data are affected by several problems; on one hand, a very large number of zero values are recorded, and on the other hand the rain gage precision (about 0.2 mm) implies an almost discrete measurement of cumulated rain as shown in Table 19.1.

19.2 The Model

Let $X(t, p)$ be the latent rainfall field at cell p and time t. $L_{t,p}$ denotes the number of lightnings in cell p at time t. Given the partially discrete nature of the dataset, following [5, 6], we discretize the latent process $X(t, p)$ below 1 mm assuming that there exist five values $\lambda_i, i = 0, \ldots, 4$ described in Table 19.1, that occur with positive probability whenever $X(t, p)$ belongs to one of the interval reported in the same table.

Let $Y(t, p)$ be the latent rainfall field on the log scale. $Y(t, p)$ is modeled as the sum of a fixed effect and a space-time random effect **W**:

$$y(t, p) = \mu(t, p) + w(t, p) + \epsilon(t, p), \tag{1}$$

where $\mu(t, p)$ is as in Eq. 3 and $w(t, p)$ is the (t, p) element of **W**, a separable space-time random field such that $w(t, p) = T(t) + S(p)$ with $T(t) = \alpha T(t-1) + \eta(t)$, $\eta(t) \sim N(0, \sigma_\eta^2)$, and

$$\mathbf{S} \sim MN(\mathbf{0}, \sigma_s^2(\mathbf{I} - \rho_s \mathbf{B})^{-1}), \tag{2}$$

where **I** is the identity matrix and **B** an adjacency matrix describing the spatial neighborhood structure. $\epsilon(t, p) \sim N(0, \tau^2)$ are independent, identically distributed random variables.

19.2.1 The Fixed Effect

The fixed component of the model relates precipitations and lightnings starting from the well known Tapia–Smith–Dixon relation [7]. More precisely, we assume that the number of lightnings in cell p depends on the number of lightnings occurring in neighboring cells. We adopt a queen neighboring structure [3], and $\omega_{i,p} = \frac{L_{i,p}}{L_p} + \frac{1}{8}\frac{L_{i,N_p}}{L_p}$ are the corresponding spatial weights where $L_{i,p}$ and $L_{i,N_p} = \sum_{P_s \in N_p} L_{i,P_s}$ are the number of lightnings in predicting cell p and in its neighborhood at time i, respectively, and $L_p = \sum_{i=1}^{T}\left(L_{i,p} + L_{i,N_p}\right)$. Moreover, we assume that the number of lightnings at time t is a function of storm propagation speed V with two different parametrization depending on the stage of the event. In fact, the life of lightning pattern inside a rainfall convective event is composed of three stages: *Charging phase* (Ch), *Mature state* (Ma), and *Dissipating phase* (Dis). Consequently, the event duration interval can be partitioned into $[t_0, T_{\text{Ch}})$, $[T_{\text{Ch}}, T_{\text{Ma}})$, and $[T_{\text{Ma}}, T]$. Thus, the fixed effect can be described as follows:

$$\mu(t,p) = \log\left(C * \sum_{i=1}^{T} L_{i,p} * \left(\exp\left\{-\frac{(a+bV)}{A_p^{1/2}}|t - T_i|^2\right\}I_{[T_{\text{Ch}},T]}(t)\right.\right.$$
$$\left.\left. + \exp\left\{-\frac{(a+bV)}{A_p^{1/2}}|t - T_i|\right\}I_{[0,T_{\text{Ch}}]}(t)\right) + C * \sum_{i=1}^{T} \omega_{i,p}\right), \quad (3)$$

where C is a mass-to-volume conversion factor which is linked to the rainfall lightning ratio [7], $I_{[t',t'']}(\cdot)$ is the indicator function of the time interval $[t',t'']$, and A_p is the area of a single cell. Here, t is a general time point and T_i is the observed time.

19.2.2 The Spatial Component

The spatial component S is modeled using a conditional autoregressive model (CAR) [2]. Let $D = \{1, \ldots, p, \ldots, n\}$ be the spatial domain, N_p the neighborhood of cell p such that $N_p \equiv \{p^* \in D : p^* \neq p \text{ is a neighbor of } p\}$ $p, p^* \in D$, and $S^t = (S_1, \ldots, S_p, \ldots, S_n)$ the space random field at each time t. For the sake of simplicity, let us omit the t notation. The CAR model is a Markov random field, then the knowledge of the set of conditional distributions identifies the joint distribution (under very general conditions); furthermore, conditional distributions depend only on the neighborhood structure. Details can be found for example in [1,3] or [2].

Our choice of neighborhood structure is a second-order nearest-neighbor structure where at least $k = 2$ neighbors are selected (the resulting graph is in Fig. 19.1b). The maximum number of neighbors is 4. We randomly select 88 cells for estimation and 23 cells for validation purposes.

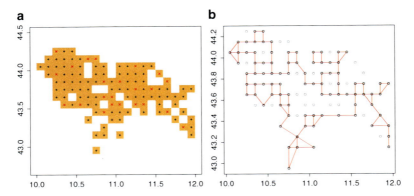

Fig. 19.1 Spatial domain. (**a**) the whole 111 cells (*orange*) and 88 selected cells for estimation (*black crosses*). (**b**) neighborhood graph of the estimation cells. The remaining 23 cells in panel (a) (*red crosses*) as well as *light grey points* in panel (b) are the selected validation cells

19.3 Estimation of Parameters and Preliminary Results

The hierarchical formulation of the model derives from Eq. 1 conditioning the latent variable on the space-time random process $W = \big(w(t_1, p_1), \ldots, w(t_T, p_n)\big)^T$ such that

$$\mathbf{Y}|\theta, \mathbf{W} \sim MN(\mu + \mathbf{w}, \tau^2 I). \qquad (4)$$

The complete set of parameters of our model is $\theta = \{a, b, \alpha, \tau_\eta^2, \tau_S^2, \rho_S, \tau^2\}$ where $\sigma_\eta^2 = 1/\tau_\eta^2$ and $\sigma_S^2 = 1/\tau_S^2$. Assuming that $\mathbf{W}$ is a stationary Gaussian process, the $Y_{t,p}$ are conditionally independent Gaussian random variable given $\mathbf{W}$ and follow a linear mixed model with expectations $E(Y_{t,p}|W) = \mu_{t,p} + w_{t,p}$ and common variance τ^2. Thus, we formulate the data likelihood conditional on the Gaussian spatiotemporal random effects $\mathbf{W}$ at level 1; we define the space-time auto-regressive structure at level 2, and we choose the priors at level 3:

Level 1 $\mathbf{Y}|\theta, \mathbf{W} \sim MN(\mu + \mathbf{w}, \tau^2 I)$

Level 2 $w = T + S;\ T(t)|\alpha, \tau_\eta^2 \sim N(\alpha T(t-1), \frac{\sigma_\eta^2}{1-\alpha^2});\ \mathbf{S}|\tau_S^2, \rho_S \sim MN(\mathbf{0}, \sigma_s^2(\mathbf{I} - \rho_s\mathbf{B})^{-1});\ \mu = f(\mathbf{L}, a\ b; V, T_{\text{Ch}})$

Level 3 $a \sim \Gamma(a_0, b_0),\ b \sim \Gamma(a_1, b_1);\ \alpha \sim N(\mu_\alpha, \sigma_\alpha^2),;\ \tau_\eta^2 \sim \text{Inv}\Gamma(a_\eta, b_\eta);\ \tau_S^2 \sim \text{Inv}\Gamma(a_S, b_S);\ \rho_S \sim N(0, \sigma_\rho^2)I_{(0,1/4)};\ \tau^2 \sim \text{Inv}\Gamma(a_\tau, b_\tau)$

The model is implemented in JAGS [4] using *R2jags* [8] to run the simulation within R. The MCMC converges rapidly; we run two chains with dispersed starting points for 20,000 iterations, with a burn-in of 5,000, and we retain the last 1,000 iterations of each chain for estimation. Convergence was inspected both graphically and from several statistics. Simulations summaries are reported in Table 19.2.

Table 19.2 Simulation summaries and posterior inference

Parameters	Mean	sd	2.5%	25%	50%	75%	97.5%	$\hat{R}$	n.eff
a	0.00	0.00	0.00	0.00	0.00	0.00	0.00	1.26	13.00
α	1.00	0.01	0.97	0.99	1.00	1.01	1.02	1.00	1000.00
b	106.73	10.25	87.94	99.29	106.48	113.24	128.83	1.00	1000.00
ρ_S	0.08	0.05	0.00	0.03	0.07	0.11	0.19	1.00	1000.00
τ_η^2	52.79	6.85	40.34	48.16	52.57	57.33	67.37	1.00	1000.00
τ_S^2	16.23	2.37	11.93	14.62	16.14	17.81	20.81	1.00	620.00
τ^2	10.19	0.19	9.79	10.08	10.19	10.30	10.55	1.00	1000.00

From this first series of simulations we compute, as a preliminary check, predicted values at validation sites as posterior plug-in estimates from the model. We check prediction quality by RMSE. Results are very encouraging as the obtained RMSE is around 1 mm.

References

1. Banerjee S, Carlin BP, Gelfand AE (2004) Hierarchical modeling and analysis for spatial data. Chapman and Hall/CRC, Boca Raton
2. Besag JE (1974) Spatial interaction and the statistical analysis of lattice systems. J R Stat Soc Series B 36:192–236
3. Cressie NAC (1993) Statistics for spatial data, revised edn. Wiley-Interscience, New York
4. Plummer M (2003) JAGS: a program for analysis of bayesian graphical models using gibbs sampling. In: Proceedings of the 3rd international workshop on distributed statistical computing (DSC 2003), Vienna, 20–22 Mar 2003. ISSN 1609-395X
5. Jona Lasinio G, Sahu SK, Mardia KV (2007) Modeling rainfall data using a Bayesian Kriged-Kalman model. In: Bayesian statistics and its application. Anamaya Publisher, New Delhi, pp 301–318
6. Sahu SK, Jona Lasinio G, Orasi A, Mardia KV (2005) A comparison of spatio-temporal Bayesian model for reconstruction of rainfall fields in cloud seeding experiment. J Math Stat 1(4):272–280
7. Tapia A, Smith JA, Dixon M (1998) Estimation of convective rainfall from lightning observations. J Appl Meteor 37:1497–1509
8. Yu-Sung S, Masanao Y (2012) R2jags: a package for running jags from R. R package version 0.03-08. http://CRAN.R-project.org/package=R2jags

Chapter 20
Bayesian Approach to Environmental Problem Based on PFLOTRAN Package

Orest Dorosh, Henryk Wojciechowicz, and Piotr Kopka

Abstract Presented result of applying Bayes inference for searching an isolated source of contamination in soil using program PFLOTRAN.

20.1 Introduction

Due to the progress in the development of computing clusters and computational algorithms quite widely is used dedicated software for modelling problems related to the transport and dispersion of contamination in the environment [1], allowing for simulation physical processes in big domains and high-resolution grids.

In particular, for the dispersion of contamination in the porous media, we have used program PFLOTRAN, a massively parallel 3-D reservoir simulator [2] (http://ees.lanl.gov/pflotran/), developed at LANL/ORNL, USA. PFLOTRAN can model multiphase reactive flows in geologic formations based on continuum scale mass and energy conservation equations. It employs the PETSc (Portable, Extensible Toolkit for Scientific Computation), a numerical modular package and efficient Newton–Krylov solver framework.

We present here a computational problem related to the identification of source of contamination in porous media, by solving inverse problem, basing on the Bayesian approach.

O. Dorosh (✉) • H. Wojciechowicz • P. Kopka
National Centre for Nuclear Research, Świerk-Otwock, Poland
e-mail: orest.dorosh@ncbj.gov.pl; henryk.wojciechowicz@ncbj.gov.pl; piotr.kopka@ncbj.gov.pl

20.2 Problem

Our goal is to localize source of contamination in soil with underground water flow using information from data of set of sensors. For this inverse problem we use statistic method [3] coupled with software package PFLOTRAN.

The area of calculation is a rectangular parallelepiped of size $5,000 \times 2,500 \times 100$ m. For our inverse problem we generated synthetic data by PFLOTRAN. Then statistic methods are applied for localization of source comparing data in sensor position.

20.3 Bayesian Approach

As "experimental data" we used the results of calculations with position of the well (the contamination source) at place $(x, y) = (1050\,\text{m}, 1350\,\text{m})$. The map of contamination concentration is presented on Fig. 20.1a. The contamination is spread from lower part of well in depth from 20 to 65 m below level of ground. The values of contaminate concentration taken in eight different points of calculation domain are considered as "experimental data" (in Fig. 20.1, a sensors are marked as crosses).

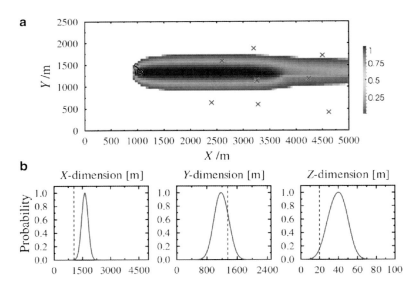

Fig. 20.1 (a) XY plane of calculation domain. **S** is a position of actual source of contamination. The *crosses* are the position of the sensors. The scale of relative concentration of contaminants. (b) Three components of posterior distribution. The *dashed lines*—actual values of source position components

The algorithm of source searching consists of the following steps:

1. For prior distribution choose set of possible well positions. The probability for choosing the position of well is proportional to prior probability distribution for the position.
2. Perform simulation with PFLOTRAN for each of well positions.
3. Compare the data from sensors set with the "experimental data" in "sensors" location, and take some subset (ten in our case) locations of sources that have lowest value of fitting function.
4. Update and normalize posterior distribution and use it as a prior distribution for the next stage.

For every stage we performed 50 simulations with different locations of contamination source. With fitting function $f = \sum_{i=1}^{8}(\eta_{\text{obs.}}^i - \eta_{\text{calc.}}^i)^2$ (where η is a concentration of contaminants and summation is by sensors) we choose ten source locations to calculate posterior distribution, according to the formula:

$$\rho_{\text{posteriors}}(\mathbf{r}) = \sum_{i=1}^{10} \rho_{\text{prior}}(\mathbf{r}) \cdot \exp\left(-\frac{(x-x_i)^2}{2\sigma_x^2} - \frac{(y-y_i)^2}{2\sigma_y^2} - \frac{(z-z_i)^2}{2\sigma_z^2}\right), \quad (1)$$

where (x_i, y_i, z_i) is the location of source from ten best. The variances are: $\sigma_x = 500$ m, $\sigma_y = 500$ m, and $\sigma_x = 10$ m. For the first stage we used uniform prior distribution.

20.4 Results and Discussion

The resulting distribution (after 11 iteration steps) gives the most probable values for source position $(x, y, z) = (1640\,\text{m}, 1180\,\text{m}, 40\,\text{m})$ (Fig. 20.1b). The difference with actual source location $\Delta \mathbf{r} = (590\,\text{m}, -170\,\text{m}, 20\,\text{m})$. The big difference for the X component can be explained by the symmetry of the problem. Some sensors are out of the contamination flow (Fig. 20.1a) and do not contribute in the process of the distribution recalculation. But information from simulation we performed can be helpful in putting sensors in new places for better collecting measurement data.

References

1. Gousseaua P, Blockena B, Stathopoulosb T, van Heijst GJF (2011) CFD simulation of near-field pollutant dispersion on a high-resolution grid: a case study by LES and RANS for a building group in downtown Montreal. Atmos Environ 45:428–438
2. Lichtner PC, Hammond G (2011) Quick Reference Guide: PFLOTRAN 2.0 (LA-CC-09-047). Multiphase-multicomponent-multiscale massively parallel reactive transport code. 29 Mar 2011
3. Thomsona LC, Hirst B, Gibsona G, Gillespie S, Jonathan Ph, Skeldona KD, Padgett MJ (2007) An improved algorithm for locating a gas source using inverse methods. Atmos Environ 41:1128–1134

Chapter 21
Bayesian Hierarchical Modeling of Growth via Gompertz Model: An Application in Poultry

Emre Karaman, Ebru Kaya, Dogan Narinc, and Mehmet Z. Firat

Abstract Estimation of growth curves of poultry species is of particular importance in animal science. This study aims at fitting hierarchical Gompertz growth curve to Japanese quail's body weight data obtained from hatching to 56 days of age weekly. The model involved a random animal effect as well as sex and line effects in asymptotic weight parameters. Model parameters were estimated via Bayesian methodology. The results of the present study indicated a higher asymptotic weight for selection line quail than that of control line. Moreover, as expected, female quails had a higher asymptotic weight than males.

21.1 Introduction

To date, the use of mixed models, which include both fixed (e.g., sex, line) and random (individual) effects, is scarce in modeling the growth of poultry [1–3]. It is also natural to assign a hierarchy to this type of data, such that the response variable body weights can be regarded as nested within individuals [4]. Although Bayesian analysis of hierarchical linear models was considered by various researchers [5, 6], in fitting nonlinear models, particularly to growth data of poultry, the use of Bayesian approach is rare [7]. In the present study, widely used Gompertz growth model was fitted to Japanese quail's data using Bayesian approach.

E. Karaman (✉) • E. Kaya • D. Narinc • M.Z. Firat
Department of Animal Science, Akdeniz University, Antalya, Turkey
e-mail: emrekaraman@akdeniz.edu.tr; ebrukaya@akdeniz.edu.tr; dnarinc@akdeniz.edu.tr; mzfirat@akdeniz.edu.tr

21.2 Material and Method

Weekly body weight measurements of 206 male and female Japanese quail from hatching to 56 days of age were formed the basis of this study. Data involves the measurements of a selected line, which was obtained by selecting the quail for 4th-week high body weight over 4 generations, and a control line.

Consider the following fixed effects Gompertz growth model:
$$y_j = \beta_0 \exp[-\beta_1 \exp(-\beta_2 t_j)],$$
where y_j is the body weight at time t_j, β_0 is the final (asymptotic) weight, β_1 is the time-scale parameter, and β_2 is the growth rate. In this function, any combination of the parameters can be assumed as random or fixed. Assume a Gompertz model where only the asymptotic weight, β_0, is allowed to vary among birds in each line and sex group. The model, in terms of fixed and random effects, can now be written as
$$y_{ij} = [\beta_0 + g_1(\text{line})_i + g_2(\text{sex})_i + u_i] \exp(-\beta_1 \exp(\beta_2 t_{ij})).$$
Here, the fixed effect parameters to be estimated are $\beta_0, \beta_1, \beta_2, g_1$, and g_2, whereas the random effect to be estimated is $u_i \sim N(0, \sigma_u^2)$, the random animal deviation.

Assuming independence among individuals, the conditional distribution of the data vector **y** given the other parameters is as follows:
$$f(\mathbf{y}|.) = \prod_{i=1}^{N} \prod_{j=1}^{n_i} \left(\frac{1}{\sqrt{(2\pi)}\sigma_e} \times \exp\left\{ -\frac{[y_{ij} - [(\beta_0 + g_1(\text{line})_i + g_2(\text{sex})_i + u_i)\exp(-\beta_1 \exp(-\beta_2 t_{ij}))]]^2}{2\sigma_e^2} \right\} \right).$$

Following prior distributions were assigned to the model parameters:
$\beta_0 \sim N(0, 0.0001), \beta_1 \sim N(0, 0.0001), \beta_2 \sim N(0, 0.0001)$
$g_1 \sim N(0, 0.0001), g_2 \sim N(0, 0.0001)$
$\sigma_u^2 = \frac{1}{\tau_u}$ and $\tau_u \sim U(0, 1000)$
$\sigma_e^2 = \frac{1}{\tau_e}$ and $\tau_e \sim U(0, 1000)$.

Three chains of 100,000 cycles were considered with a burn-in period of 10,000 for each. Thinning intervals were set to 270 cycles.

21.3 Results

Estimates of asymptotic weight parameter for each sex and line group are plotted in (Fig. 21.1). It can be clearly seen from the figure that the asymptotic weight parameter of Gompertz model was affected by sex and line. Asymptotic weight parameter of females was slightly higher than that of males while the selection line quail also had a higher parameter estimate than that of control line. According to the results of the present study, it can be concluded that selection for the 4th-week high body weight produces quail with an improved asymptotic (maximum, potential) body weights.

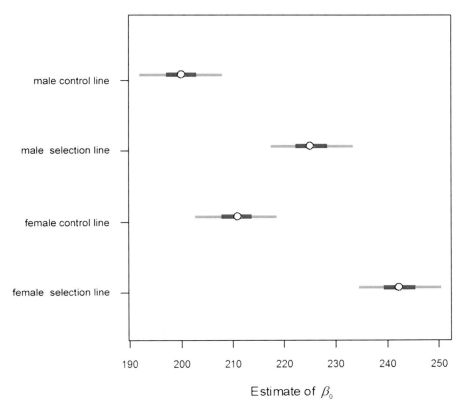

Fig. 21.1 Estimates of asymptotic weight parameter for each sex and line group

References

1. Wang Z, Zuidhof MJ (2004) Estimation of growth parameters using a nonlinear mixed Gompertz model. Poult Sci 83:847–852
2. Kizilkaya K, Balcioglu MS, Karabag K, Genc IH (2006) Growth curve analysis using nonlinear mixed model in divergently selected Japanese quails. Archiv fr Geflgelkunde 70(4):181–186
3. Aggrey SE (2009) Logistic nonlinear mixed effects model for estimating growth parameters. Poult Sci 88:276–280
4. Hall DB, Clutter M (2004) Multivariate multilevel nonlinear mixed effects models for timber yield predictions. Biometrics 60:16–24
5. Firat MZ, Theobald CM, Thompson R (1997) Univariate analysis of test day milk yields of British Holstein Friesian heifers using Gibbs sampling. Acta Agric Scand A Anim Sci 47: 213–220
6. Gelman A (2006) Prior distributions for variance parameters in hierarchical models. Bayesian Anal 1(3):515–533
7. Blasco A, Miriam Piles M, Varona L (2003) A Bayesian analysis of the effect of selection for growth rate on growth curves in rabbits. Genet Sel Evol 35:21–41

Chapter 22
Bayesian Prediction of SMART Power Semiconductor Lifetime with Bayesian Networks

Kathrin Plankensteiner, Olivia Bluder, and Jürgen Pilz

Abstract In this paper Bayesian networks are used to predict complex semiconductor lifetime data. The data of interest is a mixture of two lognormal distributed heteroscedastic components where data is right censored.

To understand the complex behavior of data corresponding to each mixture component, interactions between geometric designs, material properties, and physical parameters of the semiconductor device under test are modeled by a Bayesian network. For the network's structure and parameter learning the statistical toolboxes *BNT* and *bayesf Version 2.0* for MATLAB have been extended. Due to censored observations MCMC simulations are necessary to determine the posterior density distribution and evaluate the network's structure. For the model selection and evaluation goodness of fit criteria such as marginal likelihoods, Bayes factors, predictive density distributions, and sums of squared errors are used.

The results indicate that the application of Bayesian networks to semiconductor data provides useful information about the behavior of devices as well as a reliable alternative to currently applied methods.

22.1 Introduction

In automotive industry, end-of-life tests are necessary to verify that semiconductor products operate reliable. To save resources, accelerated stress tests [6] in combination with statistical models are commonly applied to predict the lifetime measured in

K. Plankensteiner (✉) • O. Bluder
KAI - Kompetenzzentrum Automobil- und Industrieelektronik GmbH, Europastrasse 8, A-9524 Villach, Austria
e-mail: kathrin.plankensteiner@k-ai.at

K. Plankensteiner • J. Pilz
Alpen-Adria-Universität Klagenfurt, Universitätsstrasse 65-67, A-9020 Klagenfurt, Austria

cycles to failure (CTF). Previous investigations [1, 10] have shown that the currently applied Bayesian Mixtures-of-Experts extended Coffin–Manson (MoE) model is sufficient for interpolation. In case of extrapolation, it cannot describe the complex behavior of the data and leads to inaccurate results. It is assumed that this lack of accuracy may be based on the fact that the model does not include physical parameters reflecting interactions between different geometric designs or material properties of the device under test (DUT) [10]. Hence, a Bayesian network including these factors is proposed.

22.2 Data Characteristics and Available Information

For this paper lifetime data obtained under different electrical and thermal stress conditions from a cycle stress test system [6] is investigated.

The stress test conditions are thereby defined by current (I), pulse length (t_p), repetition time (t_{rep}), and the device-specific voltage (V). Additionally, parameters considering the geometric design of the device, e.g., current density (J), are available. The main reasons for the failing of the devices are electrothermal and thermomechanical effects caused by repetitive stress. To capture these effects, temperature simulations as well as thermal and mechanical stress parameters [3, 13] are included. Altogether 18 covariates for the Bayesian network are available.

Since lifetime data is a mixture of two lognormal distributed components representing two different failure mechanisms [2], the dataset is divided into two subsets. For the model 169 and 867 data points for the first and second component, respectively, tested under 65 different stress conditions are used. Both components include censored data.

22.3 Model Development and Evaluation Results

For modeling lifetime data, Bayesian networks (BN) [7, 8, 12] are used. The nodes were assigned to be either discrete or continuous. To define the conditional probability distributions (CPDs), root and gaussian nodes are applied [9].

Different approaches using the automatic relevance determination (ARD) algorithm [11] (see Fig. 22.1) and priors on edges are investigated, because it is assumed that the number of data is too small for such a large network. The marginal likelihood is approximated with the method proposed by Draper [4]. The largest marginal likelihood value indicates the best BN for the first and second component, which is then used for lifetime modeling.

Bayesian structure and parameter learning is performed in MATLAB using an enhance version of the BNT [9] and extended MoE toolbox [1, 5] with MCMC methods. For the simulation of the posterior density distribution of the parameters normal and inverse-Gamma priors are applied.

22 Bayesian Prediction of SMART Power Semiconductor Lifetime with... 111

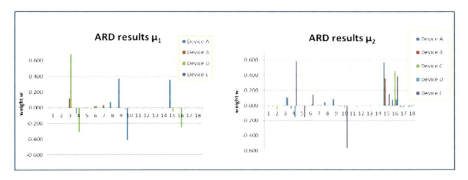

Fig. 22.1 ARD selected covariates for each component. The application of different covariates for different components is proposed

Table 22.1 Mean SSEPs

Model	Device A	Device B	Device C	Device D	Device E
# Tests	33	3	5	4	6
MoE	6.08	3.88	0.89	2.44	0.52
BN	3.83	4.12	0.96	5.26	0.22

For an evaluation, cross-validation using posterior predictive distributions and sum of squared errors of predictions (SSEPs) are compared. Furthermore, predicted outcomes are compared with the results gained by the currently applied MoE model. Since the MoE model was developed based on a subset of data, the same subset is used to learn the BNs and, thus, to provide a direct comparison between the predictive power of the two different approaches. Predicting the lifetime with BNs, the posterior predictive distribution for each component is sampled independently and mixed by estimated mixture weights. It was shown [10] that the mixture weights can be modeled by a cumulative beta distribution function.

Since it is infeasible to determine SSEPs for tests with no fails, they are neglected for this evaluation. Thus, the number of tests is reduced to 51. Table 22.1 shows the mean SSEP for the five evaluated device types. Overall, the MoE model achieves a mean SSEP of 2.76, whereas the BN model gives a mean SSEP of 2.88. The MoE model is slightly more accurate for two device types (B and C) and significantly more accurate for device type D. The BN gives significantly better results for device types A and E. Overall, the results of the MoE and BN model are comparable.

22.4 Summary

In this paper different BN have been proposed to model mixture distributed semiconductor lifetime data and to provide a reliable alternative to currently applied methods. For the model 18 covariates were available and the network's structure was supposed to be too complex for the amount of data. Therefore, the model complexity had to be reduced. This was achieved by the ARD algorithm, which provided plausible results. Furthermore, prior knowledge for edges was available which was additionally used for the structure learning.

Based on the selected network, the posterior distributions of the model parameters were simulated. The posterior densities of the model parameters show small variations and indicate therefore a good fit.

Since the aim of this work was to provide reliable predictions, cross-validation using posterior predictive distributions has been performed and evaluated. The results show that the application of a BN represents a reliable alternative to currently applied methods.

Acknowledgements The authors would like to thank Roland Sleik and Michael Ebner for the measurement support as well as Michael Glavanovics, Michael Nelhiebel, and Christoph Schreiber for valuable discussions on the topic.

This work was jointly funded by the Austrian Research Promotion Agency (FFG, Project No. 831163) and the Carinthian Economic Promotion Fund (KWF, contract KWF-1521|22741|34186).

References

1. Bluder O (2011) Prediction of smart power device lifetime based on bayesian modeling. Ph.D thesis, Alpen-Adria-Universität Klagenfurt
2. Bluder O, Pilz J, Glavanovics M, Plankensteiner K (2012) A bayesian mixture Coffin-Manson approach to predict semiconductor lifetime. SMTDA 2012: Stochastic Modeling Techniques and Data Analysis
3. Chen WT, Nelson CW (1979) Thermal stress in bonded joints. IBM J Res Dev 23(2):179–188
4. Chickering D, Heckerman D (1996) Efficient approximations for the marginal likelihood of incomplete data given a bayesian network. In: Proceedings of 12th conference on uncertainty in artificial intelligence, pp 158–168
5. Früwirth-Schnatter S (2008) Manual: MATLAB package `bayesf` Version 2.0
6. Glavanovics M, Köck H, Kosel V, Smorodin T (2007) A new cycle test system emulating inductive switching waveforms. In: Proceedings of the 12th European conference on power electronics and applications, pp 1–9
7. Hojsgaard S, Edwards D, Lauritzen S (2012) Graphical models with R. Springer, New York
8. Murphy KP (2001) An introduction to graphical models. Technical Report of UBC
9. Murphy KP. Bayes net toolbox. https://code.google.com/p/bnt.Cited15April2013
10. Plankensteiner K (2011) Application of bayesian models to predict smart power switch lifetime. Master thesis, Alpen-Adria-Universität Klagenfurt
11. Qi Y, Minka TP, Picard RW, Ghahramani Z (2004) Predictive automatic relevance determination by expectation propagation. In: Proceedings of 21st international conference on machine learning
12. Ruggeri F, Faltin F, Kenett R (2007) Bayesian networks. Encyclopedia of Statistics in quality and reliability, Wiley, New York
13. Suresh S (1998) Fatique of materials. Cambridge solid state science series. Cambridge University Press, Cambridge

Chapter 23
Consumer-Oriented New-Product Development in Fruit Flavor Breeding: A Bayesian Approach

Lebeyesus M. Tesfaye, Ivo A. van der Lans, Marco C.A.M. Bink,
Bart Gremmen, and Hans C.M. van Trijp

Abstract Taking consumer quality perceptions into account is very important for new-fruit product development in today's competitive food market. To this end, consumer-oriented quality improvement models like the quality guidance model (QGM) have been proposed. Implementing such models in the agro industry is challenging. We propose the use of Bayesian structure equation modeling (SEM) for parameterizing the quality guidance model, allowing for the integration of elicited expert knowledge. Such casual modeling would furnish important insights for determining the optimal fruit product in terms of consumer flavor-quality perceptions. In the context of tomato breeding, where we have data about metabolites, sensory-panel judgments, and consumer flavor-quality perceptions, we estimated a benchmark Bayesian SEM using non-informative priors, starting from an initial causal model derived from the data with a score-based Bayesian network (BN) learning algorithm. The results so far have given some indication of the importance of accounting for consumer heterogeneity in the modeling process.

L.M. Tesfaye (✉) • B. Gremmen
Wageningen University, Methodical Ethics and Technology Assessment,
Wageningen, The Netherlands
e-mail: Lebeyesus.Tesfaye@wur.nl; Bart.Gremmen@wur.nl

I.A. van der Lans • H.C.M. van Trijp
Wageningen University, Marketing and Consumer Behavior Group,
Wageningen, The Netherlands
e-mail: Ivo.vanderLans@wur.nl; Hans.vanTrijp@wur.nl

M.C.A.M. Bink
Wageningen University and Research Centre, Biometris, Wageningen, The Netherlands
e-mail: Marco.Bink@wur.nl

23.1 Introduction

Improving flavor-quality traits in fruit breeding calls for innovative consumer-oriented product development models. However, the wide gap in the agro sector between consumer or marketing data on the one hand side and metabolite and genomics data on the other hand side pose a great challenge to implement such modeling. In our research, we aim to link consumer flavor-quality perceptions, trained sensory-panel judgments, and various flavor-affecting metabolites in the context of tomato breeding. To address the challenge we have proposed [8] to use food-quality improvement models like the quality guidance model (QGM) of Steenkamp and Van Trijp [7] and to parameterize it using Bayesian SEM [2, 4]. This would enable us to integrate elicited expert knowledge on the degree of causal associations between metabolites and different flavor-quality perceptions in the estimation to obtain more valid and robust estimates of the strength of the different causal relations in the model. This could help flavor researchers to pin down the optimum concentration of flavor-affecting metabolites, which further can be used as phenotypes for marker association studies. Such a consumer-based process is also termed as reverse engineering. The QGM adapted for tomato-flavor quality is shown in Fig. 23.1. The perceived quality cues and attributes as well as the (overall) quality expectation and experience constitute the consumer data, while on the left we have the metabolite and the trained sensory-panel data. So far we have conducted a benchmark Bayesian SEM analysis with non-informative priors. We plan to conduct an elicitation from experts in the next stage. Section 23.2 explains the materials and methods that we used in our study.

23.2 Material and Methods

Structural equation modeling can be considered as models involving several regressions with latent variables [2], which can be represented in path diagrams. They have been used in causal theory testing and confirmation studies. SEM is a general modeling framework and hence could be estimated with different statistical methods like maximum likelihood estimation, least squares, and Bayesian methods. SEM models have essentially two key components: a measurement model that interrelates observed indicator variables (x and y indices) to underlying exogenous and endogenous latent variables (ξ and η) and a structural model that interrelates the latent variables. The regression coefficients between the indicators and the latent variables are known as factor loadings (λ). The regression coefficients between the latent variables are formally known as path coefficients (β and γ) and they represent the causal links in the model. The basic Bayesian SEM formulation is shown in Eq. 1 and is based on the formulation in Palomo et al. [4]. In Eq. 1 we have made explicit the prior representation for the latent variables:

$$p(\theta, \xi, \eta | x, y) \propto L(x, y, \xi, \eta | \theta) p(\theta; \xi; \eta) \qquad (1)$$

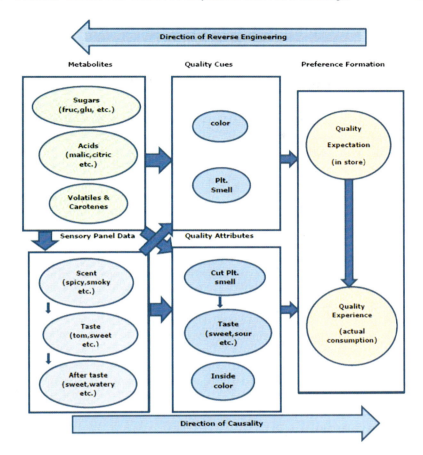

Fig. 23.1 The QGM for tomato-flavor improvement

In Eq. 1, θ is a vector that includes all the SEM model parameters, i.e., λ, β, γ, and other parameters like error terms in the measurement and structural models and variance and covariance of the latent variables. $p(\theta; \xi; \eta)$ represents the prior distributions for θ and the latent variables (ξ and η). $L(y, x, \xi, \eta|\theta)$ is the likelihood which combines with the prior to give the posterior $p(\theta, \xi, \eta|x, y)$. It has to be noted that this formulation allows the inferences of the latent variable scores alongside the other SEM model parameters [2]. MCMC methods like Gibbs sampling are applied to simulate observations from the complex posterior distribution by drawing samples iteratively from the conditional distributions of the parameters and the latent variables. Further details on this Bayesian approach to SEM could be read in [2]. Note that for our objective we are mostly interested in the path coefficients (β and γ).

In our analysis, we included non-averaged consumer ratings of 54 round tomato cultivars on 7-point Likert-scale items (from *very much disagree* to *very much agree)*. The ratings were on various flavor-quality cues and attributes consisting of color, aroma, taste as well as (overall) quality expectation and experience indicators. Trained sensory-panel judgments were measured on 1–100 line scale. In our analyses, we used averaged ratings on various sensory characteristics, such as scent, taste, and aftertaste, for the same set of round tomato cultivars. From the literature and feedback from an expert, we got a short list of 31 metabolites (consisting of acids, sugars, volatiles, and some carotenes) that are suspected to affect tomato-flavor quality. Measurements of these metabolites were extracted from a metabolite profile database on the same set of tomatoes. For a more complete description of the nature of the data and how it was generated, see Van Den Heuvel et al. [9]. We standardized the scores for all variables.

Before conducting the Bayesian SEM analysis we needed to have an initial, estimable causal model. As the available findings in the literature were not sufficient for constructing such an initial model, we derived it from our data, using a score-based Bayesian networks algorithm (hill climbing with AIC network score, implemented in the R package bnlearn [5]). In the BN learning we only used the quality-cue and quality-attribute perceptions from the consumer data. The (overall) quality expectation and quality experience latent constructs were included again in the subsequent SEM model. During the BN structure learning, we allowed only causal relations in the directions of the dark blue arrows shown in Fig. 23.1.

After learning a BN structure from the data using a score-based algorithm, we needed a way to still further reduce the relations between the metabolite and sensory variables to obtain an initial, estimable Bayesian SEM. Hence we first fitted the learned network with MLE estimation provided in the bnlearn package and deleted some relations on the basis of those results. An upper (0.5) and a lower (0.25) threshold value were used on the estimated regression weights to select the to-be-included relations between the metabolites and sensory variables. Both threshold values yielded a rather large SEM models having up to 400 path coefficients for the most extensive model (larger model).

WinBugs/R2WinBugs were used in the Bayesian SEM analysis and two MCMC chains were run up to 20 K draws with adequate burn-in for each of the two models. Non-informative priors were used using the recommendation in Lee [2]. After selecting parameters so that different parts of the model are represented (out of thousands of SEM model parameters including path coefficients, error terms, and latent scores), convergence was checked for the representative parameters by both observing the trace plots as well as the Brooks–Gelman–Rubin (bgr) diagnostic [1]. The deviance information criterion (DIC) [6] was used for model comparison and the more comprehensive model (DIC $= -50{,}607$) was selected over the more restricted model (DIC $= -50{,}580$). Model adequacy for the selected model was also checked using a posterior predictive check and most of the data falls within the 95% credible interval of the posterior predictive distribution.

Table 23.1 WinBugs output showing the path coefficient estimates for consumer and sensory-panel sweet taste feature and the path coefficient estimates within the consumer data

From	To	Mean	s.d.	MC Err.	2.5 p	Median	97.5 p
Scent-sweet	Taste-sweet	0.664	0.043	0.001	0.559	0.645	0.727
Scent-smoky	Taste-sweet	−0.471	0.081	0.003	−0.628	−0.473	−0.309
2-methylbutanal	Taste-sweet	0.053	0.087	0.003	−0.121	0.055	0.222
1-penten-3-one	Taste-sweet	0.827	0.131	0.006	0.558	0.828	1.074
3-methylbutanol	Taste-sweet	−0.318	0.104	0.004	−0.524	−0.316	−0.117
2-methylbutanol	Taste-sweet	−0.387	0.102	0.004	0.189	0.386	0.592
Cis-3-hexenal	Taste-sweet	−0.808	0.089	0.004	−0.984	−0.808	−0.634
Hexanal	Taste-sweet	0.479	0.069	0.002	0.339	0.479	0.614
Trans-2-heptenal	Taste-sweet	−0.242	0.091	0.003	−0.420	−0.243	−0.059
Methylsalicylate	Taste-sweet	0.619	0.082	0.003	0.455	0.623	0.775
1-penten-3-ol	Taste-sweet	−0.267	0.081	0.003	−0.424	−0.269	−0.107
Beta-ionone	Taste-sweet	−0.115	0.059	0.001	−0.227	−0.115	0.001
Hexanol	Taste-sweet	0.142	0.055	0.001	0.035	0.142	0.254
Scent-smoky	Sweet-tasteC	0.010	0.151	0.006	−0.298	0.012	0.297
Taste-sweet	Sweet-tasteC	−0.166	0.081	0.002	−0.328	−0.165	−0.010
Aftertaste-salt	Sweet-tasteC	0.091	0.096	0.003	−0.099	0.090	0.277
Taste-tomato	Sweet-tasteC	0.171	0.103	0.004	−0.031	0.171	0.374
Aftertaste-chemical	Sweet-tasteC	0.108	0.079	0.001	−0.047	0.107	0.266
Pleasant-smell(cut)	Sweet-tasteC	0.144	0.056	0.000	0.033	0.144	0.257
2-methylbutanal	Sweet-tasteC	0.206	0.129	0.005	−0.051	0.206	0.451
1-penten-3-one	Sweet-tasteC	−0.002	0.161	0.007	−0.323	−0.016	0.310
Cis-3-hexenol	Sweet-tasteC	−0.009	0.148	0.006	−0.289	−0.011	0.281
2-izobutylthiazol	Sweet-tasteC	−0.123	0.095	0.003	−0.308	−0.124	0.064
Phenylethanol	Sweet-tasteC	0.160	0.095	0.004	−0.024	0.160	0.352
Methylsalicylate	Sweet-tasteC	0.015	0.133	0.005	−0.238	0.012	0.288
Beta-damascenone	Sweet-tasteC	−0.057	0.117	0.005	−0.285	−0.060	0.178
3-methylbutanal	Sweet-tasteC	−0.144	0.146	0.007	−0.434	−0.144	0.140
1-penten-3-ol	Sweet-tasteC	0.011	0.131	0.005	−0.241	0.011	0.268
Hexanol	Sweet-tasteC	0.030	0.133	0.005	−0.237	0.030	0.290
Aspartic-acid	Sweet-tasteC	0.227	0.141	0.006	−0.055	0.227	0.500
Glutamate	Sweet-tasteC	−0.274	0.161	0.008	−0.582	−0.275	0.051
Glucose1	Sweet-tasteC	−0.078	0.104	0.004	−0.283	−0.078	0.128
Citric-acid	Sweet-tasteC	−0.079	0.092	0.003	−0.258	−0.079	0.102
Myolnositol	Sweet-tasteC	0.204	0.095	0.002	0.021	0.204	0.392
Sucrose	Sweet-tasteC	−0.043	0.084	0.002	−0.205	−0.043	0.121
Pleasant Smell	QexpctC	0.809	0.077	0.001	0.661	0.809	0.964
colorC	QexpctC	0.283	0.043	0.000	0.198	0.284	0.367
QexpctC	QexprncC	0.369	0.050	0.000	0.269	0.37	0.469
Pleasant-smelC(cut)	Qexprnc	0.057	0.045	0.000	−0.029	0.056	0.146
Bitter-tasteC	QexprncC	−0.011	0.046	0.000	−0.099	−0.011	0.081
Sour-tasteC	QexprncC	−0.039	0.047	0.000	−0.130	−0.039	0.052
Watery-tasteC	QexprncC	−0.032	0.041	0.000	−0.113	−0.032	0.047
Fresh-tasteC	QexprncC	0.309	0.044	0.000	0.220	0.309	0.395
Sweet-tasteC	QexprncC	0.201	0.040	0.000	0.123	0.200	0.282
colorC(cut)	QexprncC	0.174	0.045	0.000	0.085	0.175	0.263

23.3 Results

The results show many significant estimates of higher magnitude not only from metabolites to sensory-panel data but also within the consumer data (i.e., from cues/attributes to quality expectation and quality experience). We also observe many significant estimates of higher magnitude within the sensory-panel data. However, we observe many nonsignificant and/or very small estimates of the paths towards the consumer quality-cue and quality-attribute perceptions (see Table 23.1). As presenting all the output is not practical, Table 23.1 shows only a small portion of the WinBugs output taking sweet taste as representative attribute. In the table, to distinguish the consumer data from the sensory-panel data, the variables from the consumer data end with the letter "C". Hence, the label sweet-tasteC refers to the sweet taste as perceived by consumers, whereas taste-sweet refers to the judgment of sweet taste by the sensory panel. The color and smell features associated with "cut" refer to evaluations made after cutting the fruit. The metabolites are given in their full names. The table also shows all the estimates of the path coefficients from consumer cues and attributes towards the quality expectation (QexpctC) and quality experience (QexprnC). Besides the mean and median, we have also included the standard deviation (s.d.), Monte Carlo error (MC Err.), and the 95% credible interval. The underlined values are nonsignificant judged from whether the credible interval includes zero [3]. Furthermore, adjusted R^2 (not shown in the table) are higher for both the sensory-panel variables and the consumer quality expectation and quality experience, while for the consumer cues and attributes they are very small. We postulated that heterogeneity among consumers was a major cause for the mostly nonsignificant and/or small magnitude path coefficients towards the consumer cue and attributes of the QGM. This was supported by an additional Bayesian SEM analysis using an averaged consumer data that showed an increase in the values of these path coefficients.

23.4 Concluding Remarks

Based on the results, in a future research, we aim to account for consumer heterogeneity using a finite mixture Bayesian SEM [2] or by first clustering the consumer data and conducting separate standard SEM models to the different consumer segments and comparing and checking whether we have increased significant and/or higher magnitude estimates. Once this yields a suitable benchmark model, we will start to specify informative priors on the basis of elicited expert knowledge.

Acknowledgements This project was financed by the Centre for BioSystems Genomics (CBSG) and Centre for Society and Genomics (CSG) in the Netherlands.

References

1. Brooks SP, Gelman A (1998) General methods for monitoring convergence of iterative simulations. J Comput Graph Stat 7:434–455
2. Lee SY (2007) Structural equation modeling: a bayesian approach. Wiley, Chichester
3. Muthen B, Asparouhov T (2012) Bayesian structural equation modeling: a more flexible representation of substantive theory. Psychol Methods 17(3):313–335
4. Palomo J, Dubson DB, Bollen K (2007) Bayesian structural equation modeling. In: Kontoghiorghes EJ (ed) Handbook of computing and statistics with applications, vol 1, pp 163–188. Elsevier, Amsterdam
5. Scutari M (2010) Learning bayesian networks with the bnlearn R package. J Stat Softw 35(3):1–22
6. Spiegelhalter DJ, Best NG, Carlin BP, Van der Linde A (2002) Bayesian measures of model complexity and fit. J R Stat Soc 64(4):583–639
7. Steenkamp J-BEM, Van Trijp HCM (1996) Quality guidance: a consumer-based approach to food quality improvement using PLS. Eur Rev Agric Econ 23:195–215
8. Tesfaye LM, Bink MCAM, Van der Lans IA, Gremmen B, Van Trijp HCM (2013) Bringing the voice of consumers into plant breeding with bayesian modeling. Euphytica 189(3):365–378
9. Van Den Heuvel T, Van Trijp HCM, Van Woerkum C, Jan Renes R, Gremmen B (2007) Linking product offering to consumer needs; inclusion of credence attributes and the influences of product features. Food Qual Prefer 18(2):296–304

Chapter 24
Bayesian Layer Counting in Ice-Cores: Reconstructing the Time Scale

J.J. Wheatley, P.G. Blackwell, N.J. Abram, and E.W. Wolff

Abstract The concentrations of various chemicals, particles and gasses in ice-cores hold a continuous record of climatic and environmental information dating back hundreds of thousands of years. In order to interpret these data we must first learn about their underlying, unobserved time scale. We present a fully Bayesian univariate approach to obtaining a marginal posterior distribution for the time of year, as well as the date, at each depth.

24.1 Introduction

Environmental signals are measured from ice-cores either by cutting them into sections—indexed by depth—and analysing the melt-water to provide a piecewise average or via continuous flow analysis (CFA); see [1]. Some high-resolution signals, those with multiple measurements per year, have annual cycles which show as quasi-periodic seasonality in the depth series. Layer counting uses this periodicity to count back in time, year by year, and is currently achieved by eye, at considerable effort; see [1]. Using a simple, flexible model for one such signal we develop a Markov chain Monte Carlo (MCMC) approach to reconstructing the

---

J.J. Wheatley (✉) • P.G. Blackwell
School of Mathematics and Statistics, The University of Sheffield, Sheffield, UK
e-mail: j.wheatley@shef.ac.uk; p.blackwell@sheffield.ac.uk

N.J. Abram
Research School of Earth Sciences, Australian National University, Acton ACT 0200, Australia
e-mail: Nerilie.Abram@anu.edu.au

E.W. Wolff
British Antarctic Survey, Cambridge, UK
e-mail: ew428@cam.ac.uk

underlying periodic process. The latent chronology is sampled directly in a way that allows the number of cycles in the reconstruction to be changed without the need for dimension-changing algorithms such as reversible jump. We allow for the dependence in observation error and the lack of stationarity by modelling means, amplitudes and errors as continuous functions of depth.

24.2 The Model

The signal is modelled as

$$x_i = \alpha_i \sin(2\pi\tau_i) + \beta_i,$$

where τ_i is the latent time scale of interest at depth $i \in (1, 2, \ldots, n)$. The reconstruction of the signal is described by the parameters: $\theta = \{\tau, \alpha, \beta\}$. Figure 24.1 shows the model fitted to a short stretch of ammonium signal, around 11 cycles, from the NGRIP ice-core (Greenland) [2], measured at 1 mm intervals via CFA. The reconstruction is shown as a dotted black line where the data is missing, elsewhere it matches exactly to the signal.

24.2.1 Priors

The elapsed times over each depth increment are independently Gamma distributed with shape ψ and rate λ,

$$\tau_i - \tau_{i-1} \sim G(\psi, \lambda),$$

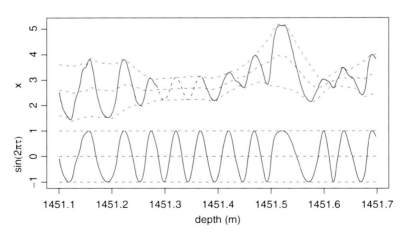

Fig. 24.1 Posterior mean reconstruction of the NGRIP ammonium signal: (*top*) the mean reconstruction, with β as a *dotted blue line* and $\beta \pm \alpha$ as a *dotted red line*; (*bottom*) $\sin(2\pi\tau)$

α and β, the amplitude and mean level of the signal, are intended to be slow-moving processes and their prior takes the form of two independent Gaussian random walks, for $i \in (2, 3, \ldots, n)$:

$$\alpha_i \sim N(\alpha_{i-1}, \sigma_\alpha^2), \quad \beta_i \sim N(\beta_{i-1}, \sigma_\beta^2).$$

24.3 MCMC Implementation

θ is updated in intervals, $I = \{i \mid s < i < f\}$, chosen uniformly at random. This is achieved using two types of Metropolis–Hastings step. The reconstruction within I is updated conditionally on that outside of I.

24.3.1 Updating θ: Maintaining the Cycle Count

The conditional distribution of α in I given α outside of I is

$$\alpha_I \mid \alpha_{-I} \sim N(\mu_\alpha, \Sigma_\alpha),$$

where

$$\mu_{\alpha,j} = \alpha_s + \frac{j(\alpha_f - \alpha_s)}{m+1}, \quad j \in (1, 2, \ldots, m),$$

and

$$\Sigma_{\alpha,jk} = \sigma_\alpha^2 \left(\min(j,k) - \frac{jk}{m+1} \right), \quad j, k \in (1, 2, \ldots, m),$$

I containing m data points, similarly for β_I. The signal in I is $\mathbf{x}_I = S\alpha_I + \beta_I$, where S is a matrix containing $\sin(2\pi\tau_I)$ along the diagonal. Thus

$$\mathbf{x}_I \mid \tau_I, \alpha_{-I}, \beta_{-I} \sim N(S\mu_\alpha + \mu_\beta, S\Sigma_\alpha S^T + \Sigma_\beta).$$

τ'_I is proposed from its prior, conditioned on τ_{-I}, by sampling $\mathbf{u}$ from a Dirichlet distribution with constant shape ψ and setting

$$\tau'_{I,j} = \tau_s + (\tau_f - \tau_s) \sum_{k=1}^{j} u_k, \quad j \in (1, 2, \ldots, m).$$

This proposal has acceptance probability

$$\frac{p(\mathbf{x}_I \mid \tau'_I, \alpha_{-I}, \beta_{-I})}{p(\mathbf{x}_I \mid \tau_I, \alpha_{-I}, \beta_{-I})}.$$

If τ_I' is accepted, S is set to S', and $(S\alpha_I)'$ is drawn from

$$S\alpha_I \mid \mathbf{x}_I, \tau_I', \alpha_{-I}, \beta_{-I} \sim N(\mu_{S\alpha}, \Sigma_{S\alpha}),$$

where

$$\mu_{S\alpha} = S\mu_\alpha + S\Sigma_\alpha S^T (S\Sigma_\alpha S^T + \Sigma_\beta)^{-1}(\mathbf{x}_I - S\mu_\alpha - \mu_\beta)$$

and

$$\Sigma_{S\alpha} = S\Sigma_\alpha S^T - S\Sigma_\alpha S^T (S\Sigma_\alpha S^T + \Sigma_\beta)^{-1} S\Sigma_\alpha S^T.$$

Then α' is set to $S^{-1}(S\alpha_I)'$ and β_I' to $\mathbf{x}_I - (S\alpha_I)'$.

24.3.2 Updating θ: Changing the Cycle Count

To add a cycle into I the proposal, τ_I', is conditioned to run between τ_s and $\tau_f + 1$. This adds a term to the acceptance probability,

$$\frac{p(\tau_I')q(\tau_I)}{p(\tau_I)q(\tau_I')} = \frac{(\Delta+1)^{(m+1)(\psi-1)}e^{-\lambda(\Delta+1)}}{\Delta^{(m+1)(\psi-1)}e^{-\lambda\Delta}},$$

where $\Delta = \tau_f - \tau_s$, which compares Δ and $\Delta + 1$ with respect to the $G((m+1)\psi, \lambda)$ distribution. If this step is accepted, τ to the right of I are incremented by 1. Cycles can be removed in a similar manner.

24.3.3 Hyper-parameters

σ_α and σ_β are given uninformative inverse-gamma priors and updated via Gibbs steps. λ is given an uninformative Gamma prior and updated via a Gibbs step. ψ is updated via a Metropolis–Hastings step with a flat, improper, prior.

24.4 Conclusions

Our approach automates the layer-counting process, providing information about the time of year, as well as the date, at each depth. The updates can be easily adapted for missing values—the reconstruction filling the gaps as seen in Fig. 24.1. A different approach to this problem can be found in [3]; the method presented here has the advantages that it is fully Bayesian and provides a more detailed chronology.

References

1. Andersen KK, Svensson A, Johnsen SJ, Rasmussen SO, Bigler M, Röthlisberger R, Ruth U, Siggaard-Andersen ML, Peder Steffensen J, Dahl-Jensen D, Vinther BM, Clausen HB (2006) The Greenland ice core chronology 2005, 15–42 ka. Part 1: constructing the time scale. Quaternary Sci Rev 25:3246–3257
2. Dahl-Jensen D, Gundestrup NS, Miller H, Watanabe O, Johnsen SJ, Steffensen JP, Clausen HB, Svensson A, Larsen LB (2002) The NorthGRIP deep drilling programme. Ann Glaciol 35:1–4
3. Wheatley JJ, Blackwell PG, Abram NJ, McConnell JR, Thomas ER, Wolff EW (2012) Automated ice-core layer-counting with strong univariate signals. Clim Past 8:1869–1879

Part IV
A Bayesian Approach to Biostatistics and Health Sciences

Chapter 25
Bayesian Analysis and Prediction of Patients' Demands for Visits in Home Care

Raffaele Argiento, Alessandra Guglielmi, Ettore Lanzarone, and Inad Nawajah

Abstract Home care (HC) providers are complex structures which include medical, paramedical, and social services delivered to patients at their domicile. High randomness affects the service delivery, mainly in terms of unplanned changes in patients' conditions, which make the amount of required visits highly uncertain. In this paper, we propose a Bayesian model to represent the HC patient's demand evolution over time and to predict the demand in future periods. Results from the application to a relevant real case validate the approach, since low prediction errors are found.

25.1 Introduction

Home care (HC) refers to any type of care provided to a patient at his/her own home. The main benefit of HC is the reduction of the hospitalization rate, which significantly increases the quality of life for the assisted patients and determines a relevant cost saving for the entire health care system [1]. Appropriate resource planning is required in HC for avoiding process inefficiencies and overloaded operators; in addition, many random events affect the service delivery and mine the feasibility of plans [5, 6]. The most frequent and critical event is a variation in patient's condition, which makes the demand for visits different from the planned one.

R. Argiento • E. Lanzarone
CNR–IMATI, Via Bassini 15, 20133 Milano, Italy
e-mail: raffaele.argiento@cnr.it; ettore.lanzarone@cnr.it

A. Guglielmi • I. Nawajah (✉)
Dipartimento di Matematica, Politecnico di Milano, Piazza Leonardo da Vinci, 32-20133 Milano, Italy
e-mail: alessandra.guglielmi@polimi.it; inad.nawajah@mail.polimi.it

In the literature, several studies deal with stochastic models for representing patient conditions in health care systems, and, among them, Bayesian approaches are also considered. As an example, Bayesian models have been used to predict patient traffic from their home to hospital in order to facilitate reconfigurations of the emergency hospital services [2]. However, to the best of our knowledge, Bayesian approaches have not been considered in the HC context so far. See [4] for a frequentist stochastic model.

The aim of this work is to propose a Bayesian model that represents and predicts the demand evolution of HC patients.

25.2 Bayesian Model

We consider m HC patients over a time period divided into discrete slots. Each patient i enters the service at time slot $T_L(i)$ and exits at $T_U(i)$. Data observed for each patient i at $t \in [T_L(i), T_U(i)]$ are:

- $N_{i,t}$: number of visits required to nurses by patient i at slot t (count data).
- $\text{CP}_{i,t}$: care profile of patient i at slot t. This is a categorical covariate (values $1, \ldots, n_s$), evolving in time, assigned by the provider based on the specific requirements and the costs associated with the provided services. Usually, a care profile is monthly confirmed or changed; however, it can be modified in advance in case of sudden variations in patient's conditions.

Moreover, patient i is characterized by sex_i (gender – categorical variable) and age_i (age in years at $t = T_L(i)$ – discrete positive variable). We model each $N_{i,t}$ as a discrete Poisson distribution with expected value $\lambda_{i,t}$. The evolution of the latent variable $\lambda_{i,t}$ over t is determined according to a Markov chain. Let $\mathbf{N_i} = \left(N_{i,T_L(i)}, N_{i,T_L(i)+1}, N_{i,T_L(i)+2}, \ldots, N_{i,T_U(i)}\right)$ for each i, and assume that $\mathbf{N_1}, \ldots, \mathbf{N_m}$ are conditionally independent. We propose the following generalized linear model (similar to [3]):

$$N_{i,t} | \lambda_{i,t} \sim \text{Pois}(\lambda_{i,t}), \qquad T_L(i) \leq t \leq T_U(i)$$

$$\log(\lambda_{i,t}) \sim N\left(\alpha[\text{CP}_{i,t}] \log(\lambda_{i,t-1}) + \beta[\text{CP}_{i,t}], \sigma^2\right), \qquad T_L(i) < t \leq T_U(i)$$

$$\log\left(\lambda_{i,T_L(i)}\right) \sim N\left(\gamma_1 \text{age}_i + \gamma_2 \text{sex}_i + \gamma_3\left[\text{CP}_{i,T_L(i)}\right], \sigma_0^2\right).$$

The latent variable $\lambda_{i,t}$ represents the health status of patient i in time slot t, which is responsible for his/her demand for visits (the bigger the parameter $\lambda_{i,t}$ is, the worse the patient's conditions are), while parameters $\alpha_s = \alpha[CP_{i,t} = s]$, $\beta_s = \beta[CP_{i,t} = s]$ and $\gamma_{3s} = \gamma_3[CP_{i,t} = s]$ describe the random effects for patient with $\text{CP}_{i,t}$. In this paper, $\text{CP}_{i,t}$ is assumed to be a fixed covariate. Parameters $\boldsymbol{\theta} = (\boldsymbol{\alpha}, \boldsymbol{\beta}, \gamma_1, \gamma_2, \boldsymbol{\gamma_3}, \sigma^2, \sigma_0^2)$ are a priori conditionally independent and their prior marginal densities are:

$$\alpha_s \overset{\text{iid}}{\sim} \mathcal{N}(0,\sigma_\alpha^2), \beta_s \overset{\text{iid}}{\sim} \mathcal{N}(0,\sigma_\beta^2), \gamma_{3s} \overset{\text{iid}}{\sim} \mathcal{N}(0,\sigma_{\gamma_3}^2), \qquad s=1,\ldots,n_s,$$

$$(\gamma_1,\gamma_2) \overset{\text{iid}}{\sim} \mathcal{N}_2(0,1000), \qquad \sigma^2 \sim U(0,5),$$

where σ_0^2 has a fixed value equal to three. Moreover, $\sigma_\alpha \sim U(0,5)$, $\sigma_\beta \sim U(0,2)$, and $\sigma_{\gamma_3} \sim U(0,15)$ are also independent. This formulation allows for predicting the future demand for visits by means of the predictive distribution of the number of nurse visits $N_{i,t+1}$ from patient i at time slot $t+1$:

$$\mathcal{L}(N_{i,t+1} = k | \text{covariate}, \mathbf{N_1},\ldots,\mathbf{N_m}) =$$
$$= \int \mathcal{L}(N_{i,t+1} = k | \lambda_{i,t+1}) \mathcal{L}(d\lambda_{i,t+1} | \lambda_{i,t}) \pi(d\lambda_{i,t} | \mathbf{N_1},\ldots,\mathbf{N_m}). \quad (1)$$

Then, the estimation $\hat{N}_{i,t+1}$ of the number of visits is assumed as the mode of its predictive distribution (1). This information is very important for HC decision makers, who are interested in assigning nurses to patients over a future planning horizon while taking into account patients' demand variability. The accuracy of the predictions for a set of m patients is evaluated in terms of the mean absolute error (MAE):

$$\text{MAE}_{t+1} = \frac{\sum_{i=1}^{m} |n_{i,t+1} - \hat{N}_{i,t+1}|}{m_t}, \quad (2)$$

where m_t is the number of patients in charge at week t, and $n_{i,t+1}$ is the observed number of nurse visits from patient i at time slot $t+1$.

25.3 Application to a Real Case

We apply and validate the model considering one of the largest Italian HC providers. The provider consists of three divisions and we refer to patients of the largest one. Patients are grouped in two categories (palliative and non-palliative patients), and each category includes a certain number of CPs. A total of 15 CPs are present in the provider. However, in this study, similar CPs are joined together and the number is reduced to 9 (Table 25.1). A more detailed description of the provider and the CPs is reported in [4]. The week is considered as the time slot, and 252 weeks (from 2004 to 2008) are included in the study. Moreover, we only consider patients who enter and exit the service once within the time window. In this way, the resulting dataset consists of 2,401 patients.

Table 25.1 Classification of CPs

Type of care	CPs of the provider	Our group
Extemporary care with a very low frequency of visits	1	1
	15	9
Integrated home care characterized by a medium–high care intensity (CPs are listed in increasing order of expected number of weekly visits)	2, 12	2
	3, 13	3
	4, 14	4
	5	5
	9	7
	10	8
Palliative care for terminal patients generally affected by oncological diseases (CPs are listed in increasing order of expected number of weekly visits)	6, 7, 8	6

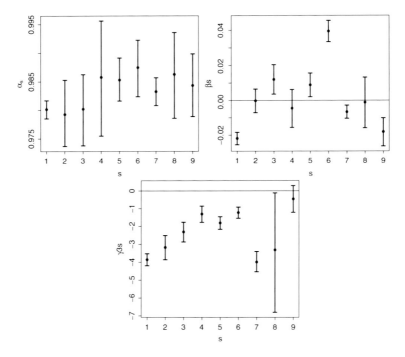

Fig. 25.1 95% credible intervals of α_s, β_s, and γ_{3s}, with $s = 1, \ldots, 9$

25.4 Results

The model was implemented in Jags [7], with 250000 iterations, burn-in equal to 5000, and thinning equal of 50. The standard convergence tests were passed. The posterior credibility intervals of the main parameters are reported in Fig. 25.1.

Table 25.2 MAE_{t+1} at 4 weeks considered for validation

$t+1$	100	150	176	235
MAE_{t+1}	0.52	0.63	0.65	0.55

In particular, the estimated values of β_s are clearly negative when CP is 1 or 9, clearly positive for CP= 6, and closer to 0 for the other CP values, leading to a subdivision coherent with the one described in Table 25.1. The validation is conducted in terms of the MAE_{t+1} at 4 different weeks $t+1$ in the observed period, i.e., $t+1=100$, $t+1=150$, $t+1=176$, and $t+1=235$. Results are reported in Table 25.2. The largest MAE is 0.65 at week 176, showing a very good fit of the model to the analyzed data and a good prediction capability.

25.5 Conclusion

In this work, we first explored the application of a Bayesian model to the HC context, in order to predict the demand for visits of the assisted patients. The approach fits well the HC context, and the results from the application to a relevant real case validate the approach, since low prediction errors are found. Hence, the applicability of the proposed model in the practice seems to be guaranteed. Future work will deal with a more complete modeling approach and, in particular, we will consider the joint estimation of the number of visits and the CPs in future periods.

References

1. Comondore V, Devereaux P, Zhou Q, et al (2009) Quality of care in for-profit and not-for-profit nursing homes: systematic review and meta-analysis. Br Med J 339:b2732
2. Congdon P (2001) The development of Gravity models for hospital patient flows under system change: a Bayesian modelling approach. Health Care Manag Sci 4:289–304
3. Giardina F, Guglielmi A, Quintana F, Ruggeri F (2001) Bayesian first order auto-regression latent variable models for multiple binary sequences. Stat Model 11:471–488
4. Lanzarone E, Matta A, Saccabarozzi G (2010) A patient stochastic model to support human resource planning in home care. Prod Plan Control 21:3–25
5. Lanzarone E, Matta A, Sahin E (2012) Operations management applied to home care services: the problem of assigning human resources to patients. IEEE Trans Syst Man Cybern A 42:1346–1363
6. Matta A, Chahed S, Sahin E, Dallery Y (2013) Modeling home care organizations from an operations management prospective. Flex Serv Manuf J. doi: 10.1007/S10696-012-9157-0
7. Plummer M (2003) JAGS: a program for analysis of Bayesian graphical models using Gibbs sampling. In: Proceedings of the 3rd international workshop on distributed statistical computing. Available via DIALOG. http://www.ci.tuwien.ac.at/Conferences/DSC-2003/Drafts/Plummer.pdf

Chapter 26
Exploiting Adaptive Bayesian Regression Shrinkage to Identify Exome Sequence Variants Associated with Gene Expression

E.M. Boggis, M. Milo, and K. Walters

Abstract Using Bayesian adaptive shrinkage in the form of the normal-gamma prior we show that causal DNA sequence variants associated with a change in gene expression can be successfully detected. Taking a fully Bayesian approach allows our model to be developed to include uncertainty in gene expression and SNP calls and to include biological information from online databases.

26.1 Introduction

Next-generation exome sequencing identifies thousands of DNA sequence variants in each individual. Methods are needed that can effectively identify which of these variants are associated with changes in gene expression, a measure of the activity of the gene. As we expect only a few SNPs (single DNA base changes) to be causal, i.e. to cause disease, we need methods that induce sparse models. The normal-gamma prior has been shown to induce adaptive shrinkage within the Bayesian linear model framework (large effects are shrunk proportionally less than small effects) [1]. Using simulated data we assess the efficacy and limitations of this Bayesian shrinkage method in comparison to other published methods in parsimoniously identifying such sequence variants. The model is then validated using publicly available human and yeast data sets. We further develop the model to include the uncertainty in gene expression; SNP functional information (information on the known biological effect of the single point mutation) obtained from online databases; and the uncertainty in the DNA base calls.

E.M. Boggis (✉) • K. Walters
School of Mathematics and Statistics, University of Sheffield, Sheffield, UK
e-mail: e.boggis@sheffield.ac.uk; k.walters@sheffield.ac.uk

M. Milo
Department of Biomedical Science, University of Sheffield, Sheffield, UK
e-mail: m.milo@sheffield.ac.uk

26.2 Modelling Using the Normal-Gamma Prior

The normal-gamma hierarchical prior [1] is given by:

$$\pi(\beta_i|\psi_i) \sim N(0, \psi_i)$$

$$\pi(\psi_i|\lambda, \gamma) \sim \text{Ga}\left(\lambda, \frac{1}{2\gamma^2}\right)$$

which has $\text{var}(\beta|\lambda, \gamma) = 2\lambda\gamma^2$ which we assign an $\text{IG}(2, M)$ prior. Consider the standard linear model

$$y_{ij} = \sum_{k=1}^{p_j} \beta_{jk} x_{ijk} + \epsilon_{ij},$$

where i, j, and k represent individual, gene, and SNP, respectively, and p_j represents the number of SNPs in the model for gene j. β_{jk} is the effect size of the kth SNP in gene j.

26.2.1 Including Uncertainty in Gene Expression (y)

To account for the uncertainty in gene expression (**y**) we use a more complex error structure. We propose to decompose the error variance to include the technical variance of the gene expression due to differences in non-biological aspects and a covariance matrix of errors representing the other unmeasurable error. The technical gene expression variance can be obtained from PUMA [2]. This method of variance decomposition estimates the variability due to technical effects and to other sources, for example, environmental and epigenetic effects (changes affecting gene expression that are not related to changes in the DNA).

26.2.2 Including Uncertainty in SNP Calls (X)

Exome sequence base calls have associated Phred-based quality scores (Q) which are a function of the base calling error probability (P), where $P = 10^{\frac{-Q}{10}}$ (high Q means high certainty in allele call). We are looking to incorporate this uncertainty using the MCMC iterations with the aim of improving detection of causal SNPs with low quality scores that might otherwise be discarded if a quality score threshold is applied.

26.2.3 Including Functional Annotation Information

Functional annotation information is increasingly widely available in online databases. Novel SNPs, SNPs that have not previously been found and recorded, are not annotated, and the only information obtainable is whether the SNP is synonymous, has no change on the protein it codes for, or non-synonymous, causes a change in the protein it codes for. Using the FS score [3], which combines a given set of biological parameters on annotated SNPs into one score, we hope to be able to prioritise more likely causal SNPs.

26.3 Preliminary Results

Our preliminary results use simulated data where confounding factors are not included. In real data sets confounding factors such as age, gender and population structure will be a problem. To avoid having to incorporate confounding in our model, we use PANAMA [4] to deconfound the gene expression signal (**y**).

In the initial simulation study 8 causal β (ranging from 2 to 0.4 in magnitude) are fixed and non-causal β are simulated to have an effect sampled from a $N(0, 0.01)$ with gene expression given by a linear sum of the weighted SNPs plus a $N(0, 1)$ error.

In comparison with the least-squares estimates (see Fig. 26.1), the standard normal-gamma [1] prior with no modifications detects all truly causal SNPs at a lower false-positive rate. This is due to comparatively less differential shrinkage across all β_{jk}. Comparing with the HyperLasso [5], which enforces similar shrinkage, and piMASS [6], which uses Bayesian selection, the normal-gamma model has similar performance (Fig. 26.1).

26.4 Conclusion

Our developments to the normal-gamma prior provide a suitable framework, which has been shown via simulation, to successfully identify causal DNA sequence variants (SNPs) affecting the gene expression level. Taking a fully Bayesian approach, permitted by the normal-gamma prior, allows for the various sources of uncertainty to be incorporated in a coherent manner.

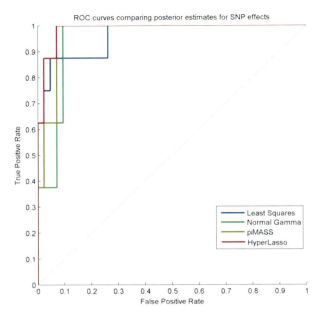

Fig. 26.1 ROC curves generated from the least squares, standard normal-gamma [1], piMASS [6], and HyperLasso [5] on simulated data

References

1. Griffin JE, Brown PJ (2010) Inference with normal-gamma prior distributions in regression problems. Bayesian Anal 5(1):171–188
2. Liu X, Milo M, Lawrence ND, Rattray M (2005) A tractable probabilistic model for Affymetrix probe-level analysis across multiple chips. Bioinformatics 21(18):3637–3644
3. Lee PH, Shatkay H (2009) An integrative scoring system for ranking SNPs by their potential deleterious effects. Bioinformatics 25(8):1048–1055
4. Fusi N, Stegle O, Lawrence ND (2012) Joint modelling of confounding factors and prominent genetic regulators provides increased accuracy in genetical genomics studies. PLoS Comput Biol 8(1):e1002330
5. Hoggart CJ, Whittaker JC, De Iorio M, Balding DJ (2008) Simultaneous analysis of all snps in genome-wide and re-sequencing association studies. PLoS Genet 4(7): e1000130
6. Guan Y, Stephens M (2011) Bayesian variable selection regression for genome-wide association studies and other large-scale problems. Ann Appl Stat 5(3):1780–1815

Chapter 27
Randomized Phase II Trials: A Bayesian Two-Stage Design

Matteo Cellamare, Valeria Sambucini, and Federica Siena

Abstract Single-arm two-stage designs are commonly used in phase II of clinical trials. However, the use of randomization in phase II trials is currently increasing. We propose a randomized version of a Bayesian two-stage design due to Tan and Machin [4]. The idea is to select the two-stage sample sizes by ensuring a large posterior probability that the true response rate of the experimental treatment exceeds that of the standard agent, assuming that the experimental treatment is actually more effective. This optimistic assumption is realized by fixing virtual outcomes.

27.1 Introduction

Phase II trials are typically conducted as single-arm studies based on a binary endpoint, where the patients are recruited in two stages to let the trial stop if the observed response rate is unacceptably low. In this context the most popular two-stage designs developed under a frequentist framework are due to Simon [2]. Among the Bayesian two-stage designs proposed in the literature, we in particular focus on the single threshold design (STD) presented by Tan and Machin [4].

Let us denote by p_X the unknown response probability of an experimental treatment and define the treatment promising if p_X exceeds a target of clinical interest, p^*. The STD selects the two-stage sample sizes by ensuring a large posterior probability that $p_X > p^*$, under the assumption that the observed response rate is

---

M. Cellamare (✉) • V. Sambucini
Sapienza Università di Roma, Piazzale Aldo Moro 5, Roma, Italy
e-mail: matteo.cellamare@uniroma1.it; valeria.sambucini@uniroma1.it

F. Siena
Unità Malattie Degenerative, Dipartimento di Neurologia Clinica e di Ricerca. Università degli Studi di Bari, Tricase (LE), Italy
e-mail: sienafederica@gmail.com

slightly larger than the target. The results are strongly affected by the choice of p^* that is typically defined *a priori* from historical data on the expected efficacy of the best available treatment. The use of historical response rates is one of the main criticisms moved to single-arm studies and the introduction of randomization in phase II of clinical trials is widely debated in the recent literature [5, 6].

A general scheme to conduct a randomized two-stage design is provided by Jung [1]. Let $\mathscr{X}$ and $\mathscr{Y}$ be the experimental and the standard arm, respectively. At the first stage, n_1 patients are enrolled in each arm. Let us denote by x_1 and y_1 the observed number of responders for $\mathscr{X}$ and $\mathscr{Y}$, respectively. If $x_1 - y_1 \geq a_1$, where $a_1 \in [-n_1, n_1]$, the trial continues to the second stage; otherwise it stops. At the second stage we accrue n_2 additional patients to each arm and observe the number of responders, x and y, out of the total of $n = n_1 + n_2$ patients. Then if $x - y \geq a$, where $a \in [a_1 - n_2, n]$, we proceed to phase III; otherwise the trial terminates. In particular, Jung [1] suggests to select the values (n_1, a_1, n, a) by minimizing either the maximum sample size or the expected sample size under the null hypothesis of no treatment difference, subject to prespecified restrictions on type I and type II error probabilities. These proposals represent randomized versions of the single-arm "minimax" and "optimal" designs due to Simon [2].

27.2 A Bayesian Two-Stage Design

To avoid the use of a historical control, we propose a randomized version of the STD. Let p_Y be the efficacy probability of the standard therapy. The criterion we suggest to select n_1 and n is based on the control of the posterior probability that $p_X > p_Y$, under the assumption that the observed response rate for the standard treatment is equal to the target p^*, while the one for the experimental treatment is equal to the target plus a small quantity $\varepsilon > 0$.

More formally, let us denote by $Pr(p_X > p_Y | X_1 = x_1, Y_1 = y_1)$ and $Pr(p_X > p_Y | X = x, Y = y)$ the posterior probabilities that $p_X > p_Y$ at the end of the first and the second stage, respectively. We select the smallest sample size n_1, such that

$$Pr(p_X > p_Y | X_1 = n_1(p^* + \varepsilon), Y_1 = n_1 p^*) \geq \lambda_1, \tag{1}$$

where $\lambda_1 \in (0, 1)$ is a prespecified threshold. Since the data arise from a binomial distribution, we introduce independent beta prior densities for the parameters, i.e. $\pi(p_j) = \text{Beta}(\alpha_j, \beta_j)$, for $j = X, Y$, where

$$\alpha_j = n_j^0 p_j^0 + 1 \quad \text{and} \quad \beta_j = n_j^0 (1 - p_j^0) + 1.$$

With this choice of the hyperparameters, the beta prior $\pi(p_j)$, for $j = X, Y$, has mode at p_j^0 and is based on an implicit *prior sample size*, n_j^0, such that the larger its value, the more concentrated is the prior distribution (see Sambucini [3]). As it is well known, the corresponding independent posterior densities for p_X and p_Y are still beta with updated parameters. Then, the posterior probability in (1) can be easily computed using, for instance, Monte Carlo simulation techniques.

Analogously, at the second stage, we choose the smallest n that satisfies

$$Pr(p_X > p_Y | X = n(p^* + \varepsilon), Y = np^*) \geq \lambda_2, \qquad (2)$$

for a suitable $\lambda_2 \in (0, 1)$. Once the optimal sample sizes have been determined and the trial started, following Tan and Machin [4], at the end of each stage, we compute the posterior probability of interest corresponding to the observed outcome and check whether it exceeds the prespecified threshold (λ_1 or λ_2) in order to make a go/no-go decision.

Finally, it is important to point out that statistical considerations about the irrelevance of stopping rules in Bayesian inference let us conclude that the posterior probability in (2) is not affected by the first-stage results. Moreover, the behavior of $Pr(p_X > p_Y | X_1 = n_1(p^* + \varepsilon), Y_1 = n_1 p^*)$ as a function of n_1 is the same as that of $Pr(p_X > p_Y | X = n(p^* + \varepsilon), Y = np^*)$ as a function of n, and we need to set $\lambda_2 > \lambda_1$ in order to obtain $n > n_1$. Then, since the posterior distributions involved in both criteria (1) and (2) are actually the same, in the following, we will use the first-stage notation in describing the numerical results related to both stages.

27.3 Numerical Results

The proposed design has been implemented and applied to some reasonable prior scenarios. Table 27.1 provides the optimal sample sizes for different values of p^* and λ_1, when $\varepsilon = 0.05$, and we consider informative prior distributions that express skepticism or enthusiasm about the efficacy of the experimental treatment. In particular we obtain a skeptical prior by specifying the prior modes $p_x^0 = p^* - 0.05$ and $p_Y^0 = p^* + 0.05$, while an enthusiastic prior is obtained by setting $p_x^0 = p^* + 0.05$ and $p_Y^0 = p^* - 0.05$. Different values of the prior sample sizes are also considered in order to take into account different levels of skepticism or enthusiasm expressed by the prior densities. As expected, larger values of λ_1 determine higher values for the optimal sample size. We can also note that, when we adopt skeptical prior densities

Table 27.1 Optimal sample sizes for different values of the prior sample sizes, p^* and λ_1, when $\varepsilon = 0.05$, and we elicit skeptical and enthusiastic prior distributions

p^*	Prior	$n_X^0 = n_Y^0 = 1$			$n_X^0 = n_Y^0 = 5$			$n_X^0 = n_Y^0 = 10$		
		λ_1			λ_1			λ_1		
		0.6	0.7	0.8	0.6	0.7	0.8	0.6	0.7	0.8
0.2	Skeptical	15	45	106	28	61	124	42	79	144
	Enthusiastic	8	38	98	1	25	86	1	6	69
0.3	Skeptical	17	55	131	31	72	149	46	90	170
	Enthusiastic	10	47	123	1	35	111	1	17	95
0.4	Skeptical	18	60	145	32	77	163	47	96	185
	Enthusiastic	11	52	137	1	40	125	1	23	109

Prior modes of *skeptical* prior distributions: $p_x^0 = p^* - 0.05$ and $p_Y^0 = p^* + 0.05$
Prior modes of *enthusiastic* prior distributions: $p_x^0 = p^* + 0.05$ and $p_Y^0 = p^* - 0.05$

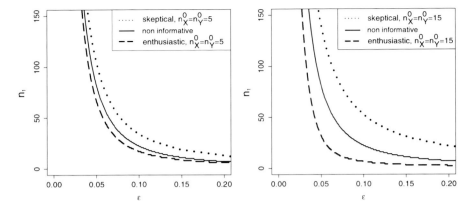

Fig. 27.1 Optimal sample size as a function of ε, when $\lambda_1 = 0.75$ and $p^* = 0.3$, using skeptical, enthusiastic and noninformative prior distributions

about the effectiveness of the new treatment, we need larger sample sizes with respect to those obtained when we use enthusiastic priors. Of course, the differences are more relevant as we increase the values of n_X^0 and n_Y^0.

Figure 27.1 represents the behavior of the optimal sample size as a function of ε, when $p^* = 0.3$ and $\lambda_1 = 0.75$, under the skeptical and the enthusiastic scenarios considered in Table 27.1 for $n_X^0 = n_Y^0 = 5$ (see left panel) and $n_X^0 = n_Y^0 = 15$ (see right panel). The case of noninformative priors is also considered by setting $n_X^0 = n_Y^0 = 0$, so that $\pi(p_j) = \text{Beta}(1, 1)$, for $j = X, Y$. As ε increases, the fixed virtual results used in the criteria (1) and (2) express a larger level of optimism about the efficacy of the experimental treatment and the design requires smaller sample sizes. Moreover, since the larger the *prior sample size*, the higher the weight assigned to the prior opinions, the difference in the optimal sample sizes under the skeptical and the enthusiastic scenarios is more evident in the right panel of Fig. 27.1.

References

1. Jung SH (2008) Randomized phase II trials with a prospective control. Stat Med 27(4):568–583
2. Simon R (1989) Optimal two-stage designs for phase II clinical trials. Controlled Clin Trials 10:1–10
3. Sambucini V (2008) A Bayesian predictive two-stage design for phase II clinical trials. Stat Med 27(8): 1199–1224
4. Tan SB, Machin D (2002) Bayesian two-stage designs for phase II clinical trials. Stat Med 21:1991–2012
5. Ratain MJ, Sargent DJ (2009) Optimising the design of phase II oncology trials: the importance of randomisation. Eur J Cancer 45(2):275–280
6. Rubinstein L, LeBlanc M, Smith MA (2011) More randomization in phase II trials: necessary but not sufficient. Natl Cancer Inst. 103(14):1075–1077

Chapter 28
Bayesian Matrix Factorization for Outlier Detection: An Application in Population Genetics

Nicolas Duforet-Frebourg and Michael G.B. Blum

Abstract We present a new Bayesian hierarchical model based on matrix factorization for detecting outliers in high-dimensional data. Outliers are explicitly modeled using both a shift-in-mean and variance inflation approach. The Bayesian framework provides intrinsic probabilities of being an outlier for each element in the sample. Posterior replicates of the parameters are simulated using a MCMC algorithm. In population genetics where many genetic markers are typed in different populations, we show that this model can be used to detect genes targeted by Darwinian selection.

28.1 Introduction

Matrix factorization aims at decomposing a high-dimensional $n \times p$ data matrix into a product of two lower rank K matrices called the factor and loading matrices [4]. Matrix factorization provides a useful framework to model outliers in the lower-dimensional space generated by the low-rank approximation [3]. Detecting outliers in high-dimensional data sets is of interest in population genetics in order to detect genes under selective pressures [1]. The proposed approach provides an intrinsic probability of being an outlier so that we can estimate false discovery rate (FDR) and q-values, which are two important quantities in whole-genome scans [6].

We provide a MCMC algorithm to sample replicates from the posterior distribution and we show how the method can detect genes under selection in population genetics data.

N. Duforet-Frebourg (✉) · M.G.B. Blum
Laboratoire TIMC-IMAG UMR 5525, Centre National de la Recherche Scientifique,
Université Joseph Fourier, Grenoble, France
e-mail: nicolas.duforet@imag.fr; michael.blum@imag.fr

28.2 Bayesian Matrix Factorization for Outlier Detection

28.2.1 Model

The probabilistic model of matrix factorization—also known as factor or probabilistic PCA model—for a design $n \times p$ matrix Y relies on a product between a factor matrix F and a loading matrix Λ:

$$Y = F\Lambda + \epsilon, \qquad (1)$$

where F is an $n \times K$ matrix, Λ is a $K \times p$ matrix, and ϵ is an $n \times p$ residual matrix where each row $\epsilon_i \sim \mathcal{N}(0_p, \sigma^2 I_p)$. Here, we choose a Gaussian prior for Λ

$$p(\Lambda|\sigma_\Lambda) = \Pi_{j=1}^p \mathcal{N}(\Lambda_j; 0_K, \sigma_\Lambda^2 I_K). \qquad (2)$$

To specify the prior of F, we explicitly model outliers using the shift-in-mean approach [5] for one of the K factors of the low-rank approximation

$$p(F|A, Z, \Sigma_F) = \Pi_{i=1}^n \mathcal{N}(F_i; 0_K + A_i^{(Z_i)}, \Sigma_F), \qquad (3)$$

where Σ_F is a diagonal matrix with values $\sigma_{F_k}^2$. We specify improper priors for variances $p(\sigma_\Lambda^2) \propto \frac{1}{\sigma_\Lambda^2}$ and $p(\sigma_{F_k}^2) \propto \frac{1}{\sigma_{F_k}^2}$. Shift vector A_is are zero-valued vectors with nonzero component at index Z_i. For $i = 1, \ldots, n$, Z_i is an integer between 0 and K, indicating that the ith line is either an outlier for the factor Z_i if $Z_i > 0$ or not an outlier if $Z_i = 0$. We add priors for A and Z such as

$$p(A|\tau, \Sigma_F) = \Pi_{i=1}^n \mathcal{N}(A_i; 0_K, \tau^2 \Sigma_F). \qquad (4)$$

$$p(Z_i = k) = \pi_k = \begin{cases} \alpha/K & \text{if } k > 0 \\ 1 - \alpha & \text{if } k = 0 \end{cases} \qquad (5)$$

where α is the expected proportion of outliers with Beta prior, $\alpha \sim Beta(\beta_1, \beta_2)$, and variance inflation parameter $\tau \sim \mathcal{U}(1, 10)$.

28.2.2 Posterior Inference and Algorithm

To obtain replicates from the posterior $p(Z, A|Y)$, we use Gibbs updating steps based on the conditional distribution of (Z_i, A_i) provided below:

$$p(Z_i = k, A_i | F_i, \Sigma_F, \tau^2) = p(Z_i = k | F_i, \Sigma_F, \tau^2) p(A_i | Z_i = k, F_i, \Sigma_F, \tau^2) \qquad (6)$$

28 Bayesian Matrix Factorization for Outlier Detection...

Table 28.1 MCMC algorithm of Bayesian Matrix Factorization for detecting outliers

• Setup values of	K, β_1, β_2.
• Initialize	$\sigma, \sigma_\Lambda, \Sigma_F, \Lambda, F, A, Z, \alpha, \tau^2$.
• for $s = 1..ns$ do:	$j = 1..p, \Lambda_j^{(s)} \leftarrow \mathcal{N}(((\frac{1}{\sigma_\Lambda^{2(s-1)}} I_K + \frac{1}{\sigma^{2(s-1)}} (F^{(s-1)})^t F^{(s-1)})^{-1}$
	$\frac{1}{\sigma^{2(s-1)}} F^{(s-1)} Y_j, (\frac{1}{\sigma^{2(s-1)}} F^{(s-1)t} F^{(s-1)} + \frac{1}{\sigma_\Lambda^{2(s-1)}} I_K)^{-1})$
	$\sigma_\Lambda^{(s)} \leftarrow IG(\frac{Kp}{2}, \frac{1}{2} \sum_{i=1}^{K} \Lambda_i^{(s)} \Lambda_i^{(s)t})$
	$i = 1..n, Z_i^{(s)} \leftarrow sample(Z_i^{(s)}, p(Z_i = k|\pi, F_i^{(s-1)}, \Sigma_F^{(s-1)}, \tau^{2(s-1)}))$
	$i = 1..n,$ if $Z_i^{(s)} > 0$ then, $A_{i,Z_i}^{(s)} \leftarrow \mathcal{N}(\frac{\tau^{2(s-1)}}{\tau^{2(s-1)}+1} F_{i,Z_i}^{(s-1)},$
	$\frac{\tau^{2(s-1)}}{\tau^{2(s-1)}+1} \sigma_{F_{Z_i}}^{2(s-1)})$
	$i = 1..n, F_i^{(s)} \leftarrow \mathcal{N}((\Sigma_F^{-1(s-1)} + \frac{1}{\sigma^{2(s-1)}} \Lambda^{(s)} \Lambda^{t(s)})^{-1}$
	$(\Sigma_F^{-1(s-1)} A_{i,Z_i}^{(s)} + \frac{1}{\sigma^{2(s-1)}} \Lambda^{(s)} Y), (\Sigma_F^{-1(s-1)}$
	$+ \frac{1}{\sigma^{2(s-1)}} \Lambda^{(s)} \Lambda^{t(s-1)})^{-1})$
	$i = 1..K, \sigma_{F_i}^{(s)} \leftarrow IG(\frac{n}{2}, \frac{1}{2} \sum_{j=1}^{n} (F_{ij}^{(s)} - A_{ij}^{(s)})(F_{ij}^{(s)} - A_{ij}^{(s)})^t)$
	$\alpha \leftarrow Beta(\beta_1 + n_{outliers}, \beta_2 + n - n_{outliers})$
	$\sigma^{(s)} \leftarrow IG(\frac{np}{2}, \frac{1}{2} \sum_{i=1}^{n} \sum_{j=1}^{p} |Y_{i,j} - F_i^{(s)} \Lambda_j^{(s)}|))$
	Metropolis-Hastings step: $\tau^{2*} \leftarrow \mathcal{N}(\tau^{2(s-1)}, .5)$

where

$$p(Z_i = k|F_i, \Sigma_F, \tau^2) \propto \pi_k \sqrt{\sigma_{F_k}^2} e^{\frac{\tau^2}{(\tau^2+1)} \frac{F_{i,k}^2}{\sigma_{F_k}^2}} \quad (7)$$

$$p(A_{i,k}|Z_i = k, F_i, \Sigma_F, \tau^2) \propto \mathcal{N}(\frac{\tau^2}{\tau^2+1} F_{i,k}, \frac{\tau^2}{\tau^2+1} \sigma_{F_k}^2) \text{ if } k > 0. \quad (8)$$

Other parameters have more usual conditional distributions that are also useful when performing Bayesian linear regression. Samples are simulated using the MCMC algorithm provided in Table 28.1.

28.3 Results

To illustrate the potential of the method, we simulate population genetics data where the outliers correspond to the markers located in genomic regions under Darwinian selection. The data contain 400 individuals from 4 populations that split according to a tree model (see Panel A in Fig. 28.1) and are typed at 200 genetic markers called SNPs, among which 12 are under various selective pressures in one of the 4 populations. Posterior probabilities to be outliers are enriched for genes targeted by selection (see Panel B in Fig. 28.1), and a Precision-Recall (see Panel C in Fig. 28.1) curve can be used to evaluate the property of the method under various evolutionary scenario.

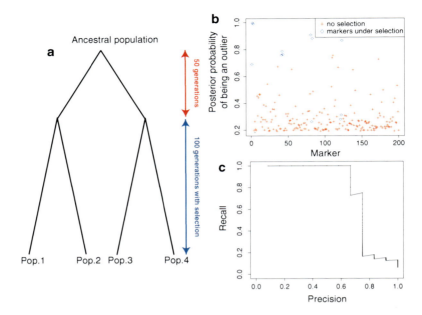

Fig. 28.1 Panel A: population divergence model used to simulate population genetics data. Panel B: posterior probability of being outlier. Panel C: Precision-Recall curve

28.4 Conclusions

We introduced a hierarchical Bayesian model of matrix factorization for detecting outliers in high-dimensional data. The probabilistic model provides probabilities for each observation to be an outlier and indicates the direction or factor under which the observation is atypical. We showed that this approach can be used to detect genes targeted by selection in population genetics. The method is implemented in the C software *PCAdapt*. The method can be more widely used in population genetics to uncover geographic direction in which selection happened since in spatially structured populations, factors are closely correlated to geography. A potential extension of the model would be to include spatial autocorrelation between individuals using model of spatial decay of correlation as in [2].

References

1. Beaumont MA, Balding, DJ (2004) Identifying adaptive genetic divergence among populations from genome scans. Mol Ecol 13(4):969–980
2. Duforet-Frebourg N, Blum MGB Non-stationary patterns of isolation-by-distance: inferring measures of local genetic differentiation with Bayesian Kriging (submitted)
3. Polasek W (1997) Factor analysis and outliers: A Bayesian approach. Wirtschaftswissenschaftliches Zentrum der Universitt Basel

4. Salakhutdinov, R, Mnih A (2008) Bayesian probabilistic matrix factorization using Markov chain Monte Carlo. In Proceedings of the 25th international conference on Machine learning pp. 880–887. ACM.
5. Verdinelli I, Wasserman L (1991) Bayesian analysis of outlier problems using the Gibbs sampler. Statist Comput 1(2):105–117
6. Wakefield J (2007) A Bayesian measure of the probability of false discovery in genetic epidemiology studies. Am j Hum Genet 81(2):208

Chapter 29
Noise Model Selection for Multichannel Diffusion-Weighted MRI

Edward Knock, Theodore Kypraios, Paul Morgan, and Stamatios Sotiropoulos

Abstract We examine the use of various diagnostics for model choice for multichannel diffusion-weighted MRI, which is important for inferring the correct tractography, as noise properties can differ between reconstruction techniques and scanners. These are calculated for image data obtained under various different settings of a Philips Achieva 3T scanner. A simulation study was carried out which showed these to be reasonably effective at identifying the true model.

29.1 Introduction

It is well known that the application of multichannel receiver arrays and new image reconstruction techniques, such as parallel imaging, can influence the noise properties in magnetic resonance imaging (MRI) [4]. It has been recently shown that different image reconstructions (which also differ between scanners) of the same multichannel raw data can significantly influence the estimation of fibre orientations from diffusion-weighted MRI [6] and therefore tractography. This is due to their different noise properties and the nature of the DW signal; the signal attenuation is of interest and therefore the signal can be very close to the noise floor. Any changes in the noise properties can directly influence the estimation process.

E. Knock (✉) • T. Kypraios
School of Mathematical Sciences, University of Nottingham, Nottingham, UK
e-mail: Edward.Knock@nottingham.ac.uk; Theodore.Kypraios@nottingham.ac.uk

P. Morgan
Medical School, University of Nottingham, Nottingham, UK
e-mail: Paul.Morgan@nottingham.ac.uk

S. Sotiropoulos
FMRIB, University of Oxford, Oxford, UK
e-mail: Stam@fmrib.ox.ac.uk

29.2 Background

29.2.1 Data

The raw data were acquired using a Stejskal-Tanner diffusion-weighted (DW) pulse sequence within a single-shot EPI (echo planar imaging) protocol. One b=0 s/mm^2 and 61 DW volumes at b=1000 s/mm^2 were acquired in a Philips Achieva 3T scanner. In-plane spatial resolution was 2x2 mm^2 and slice thickness 2mm. A similar protocol was repeated with the DW volumes being acquired at b=3000 s/mm^2. Both acquisitions were first performed using an 8-channel receiver coil and were repeated using a 32-channel coil. Magnitude images were reconstructed from the raw multichannel data in two different ways, provided by the vendor, CLEAR On (Con) and CLEAR Off (Coff). For the 8-channel and 32-channel data, respectively, these images are composed of 112x112x32 voxels and 112x112x8 voxels, from which a subset of 5914 and 2859 voxels were chosen from the midbody of the corpus callosum.

29.2.2 Models

We are interested in fitting the following ball-and-sticks model [3] for the diffusion signal for the ith acquisition:

$$S_i = S_0 \left((1 - \sum_{j=1}^{N} f_j) \exp\{-b_i d\} + \sum_{j=1}^{N} f_j \exp\{-b_i d \mathbf{g}_i^\top \mathbf{v}_j\} \right), \quad (1)$$

where the unknown parameters are: d, the diffusivity; S_0, the signal with no diffusion gradients applied; f_j ($j = 1, \ldots, N$), the proportion of signal described by fibre direction $\mathbf{v}_j$; (θ_j, ϕ_j) ($j = 1, \ldots, N$), spherical polar coordinates describing fibre direction $\mathbf{v}_j = (\sin\theta\cos\phi, \sin\theta\sin\phi, \cos\theta)$. The b-value and gradient direction associated with the ith acquisition, b_i and $\mathbf{g}_i$, respectively, are known.

This model features a mixture of tensors consisting of N perfectly anisotropic tensors (the 'sticks'), each of which depicts one fibre orientation, and a perfectly isotropic tensor (the 'ball'), which captures the rest diffusion processes (Fig. 29.1).

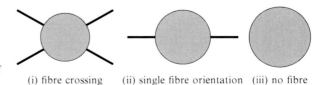

Fig. 29.1 Some examples of a ball-and-sticks model

(i) fibre crossing (ii) single fibre orientation (iii) no fibre

Here we are interested in fitting $N = 3$, in which $N = 1$ and $N = 2$ are nested by setting $f_2 = f_3 = 0$ and $f_3 = 0$, respectively.

We consider five noise models for the reconstructed signal Y_i (with their parameters):

1. **Gaussian/Normal**, $Y_i \sim \mathrm{N}\left(S_i, \tau^{-1}\right)$
2. **Rician**, $Y_i \sim \mathrm{Rice}\left(S_i, \tau\right)$
3. **Noncentral Chi** with **fixed** number of channels ($n = 8/n = 32$), $Y_i \sim \mathrm{NC}\chi_n\left(S_i, \tau\right)$
4. **Noncentral Chi** with **unknown** number of (independent) channels, $Y_i \sim \mathrm{NC}\chi_n\left(S_i, \tau\right)$
5. **Gaussian/Normal** modified by accounting for **higher noise floor**, $Y_i \sim \mathrm{N}\left(S_i, \tau^{-1}\right)$ with

$$S_i = S_0 \left(f_0 + \left(1 - \sum_{j=0}^{N} f_j\right) e^{-b_i d} + \sum_{j=1}^{N} f_j \exp\left\{-b_i d \mathbf{g}_i^\top \mathbf{v}_j\right\} \right) \quad (2)$$

29.2.3 Methods

We consider three diagnostics for model choice: Akaike information criterion (AIC) [1], Bayesian information criterion (BIC) [5] and deviance information criterion (DIC) [7]. The AIC and BIC are (maximum likelihood) point estimates, and hence can be quick to calculate, though in practice finding the true maximum may be difficult. The BIC places a higher penalty on each additional parameter. The DIC requires estimation of the likelihood at the mean and the mean of the log-likelihood. In practice this requires a full Bayesian approach using an MCMC algorithm. For speed, we use an MCMC sampler using an adaptive multivariate normal random walk proposal [8] and fit the ball-and-three-sticks model with automatic relevance determination (ARD) priors [2] proportional to $\frac{1}{f}$ on f_2 and f_3. This sparsity-inducing prior should force these to zero except when they are in truth sufficiently large. This is done to effectively fit three models at once and reduce overfitting.

29.3 Result Summary

It is found that 32 channels, b-values of 3000 and Coff reconstruction tend to favour noncentral Chi noise with degrees of freedom much lower than 32 (indicating correlation between channels), while 8 channels, b-values of 1000 and Con reconstruction favour Gaussian or Rician noise (see Table 29.1).

A simulation study carried out showed ARD priors do a good job in reducing overfitting the number of sticks though generally an ill-fitting noise model leads to

Table 29.1 Percentage of voxels for which each noise model yields lowest AIC/BIC/DIC (across all numbers of sticks)

		Normal	Rice	NCχ ($n = 8/32$)	NCχ (n unknown)	Mod. Norm.
8ch/b1k/Coff	AIC	49.98	21.98	21.03	2.65	4.35
	BIC	52.65	23.79	22.39	0.61	0.56
	DIC	21.54	24.28	10.57	21.37	22.24
8ch/b1k/Con	AIC	49.02	19.92	23.96	2.79	4.31
	BIC	51.34	21.66	25.75	0.64	0.61
	DIC	21.86	23.03	11.24	20.53	23.33
8ch/b3k/Coff	AIC	6.53	50.59	0.12	31.32	11.45
	BIC	5.83	70.46	0.30	18.26	5.14
	DIC	10.74	39.74	0.34	30.86	18.33
8ch/b3k/Con	AIC	7.73	52.42	0.57	26.90	12.38
	BIC	7.14	70.41	0.90	15.84	5.72
	DIC	10.69	39.75	0.66	28.88	20.02
32ch/b1k/Coff	AIC	50.79	27.95	14.17	2.97	4.13
	BIC	53.17	30.95	13.85	1.36	0.66
	DIC	23.05	24.38	6.44	18.96	27.18
32ch/b1k/Con	AIC	53.17	23.26	16.23	3.50	3.85
	BIC	54.56	26.27	17.10	1.29	0.77
	DIC	22.25	25.11	7.56	18.78	26.30
32ch/b3k/Coff	AIC	8.53	23.57	0.03	52.61	15.25
	BIC	8.43	36.10	0.07	45.33	10.07
	DIC	9.58	22.84	0.07	35.47	32.04
32ch/b3k/Con	AIC	13.92	38.89	0.03	33.89	13.26
	BIC	13.61	52.29	0.07	25.39	8.64
	DIC	13.05	32.00	0.28	25.88	28.79

overfitting and, hence, incorrect tractography. The diagnostics are seen to identify the true noise model reasonably well.

References

1. Akaike H (1974) A new look at the statistical model identification. IEEE Trans Automat Cont 19:716–723
2. Behrens TEJ, Johansen Berg H, Jbabdi S, Rushworth MFS, Woolrich MW (2007) Probabilistic diffusion tractography with multiple fibre orientations: What can we gain? NeuroImage 34: 144–155
3. Behrens TEJ, Woolrich MW, Jenkinson M, Johansen-Berg H, Nunes RG, Clare, S, Matthews PM, Brady JM, Smith SM (2003) characterization and propagation of uncertainty in diffusion-weighted MR imaging. Magn Reson Med 50:1077–1088
4. Dietrich O, Raya JG, Reeder SB, Ingrisch M, Reiser MF, Schoenberg SO (2008) Influence of multichannel combination, parallel imaging and other reconstruction techniques on MRI noise characteristics. Magn Reson Imag 26:754–763

5. Schwarz GE (1978) Estimating the dimension of a model. Ann Stat 6:461–464
6. Sotiropoulos SN, Behrens T, Andersson J, Yacoub E, Moeller S, Jbabd S (2011) Influence of Image Reconstruction from Multichannel Diffusion MRI on Fibre Orientation Estimation. OHBM Annual Meeting, p. 595 Wth, Quebec, Canada, June 2011. e-poster: http://humanconnectome.org/hosted/posters/HCP-HBM%202011%20poster%20595%20-%20Sotiropolous%20et%20al.pdf.Cited20May2013
7. Spiegelhalter DJ, Best NG, Carlin BP, van der Linde A (2002) Bayesian measures of model complexity and fit (with discussion). J R Stat Soc, Series B 64:583–639
8. Roberts GO, Rosentha JS (2009) Examples of Adaptive MCMC. J Comp Graph Stat 18:349–367

Chapter 30
Analysis of Hospitalizations of Patients Affected by Chronic Heart Disease

Alice Parodi, Francesca Ieva, Alessandra Guglielmi, and Raffaele Argiento

Abstract In this paper we present a Bayesian model to analyze sequences of hospitalizations of patients affected by chronic heart disease, focusing not only on the sequence but also on the times between two next events; considering covariates and time, the model is able to identify the most relevant factors influencing the evolution.

30.1 Introduction

In this paper we present a preliminary analysis of the evolution of chronic heart disease [Sect. 30.2] considering as variables of interest the sequence of hospitalizations of patients affected by this illness and the times between two next hospitalizations (see [1] for an overview on event history analysis). Data refers to patients admitted to hospitals in Regione Lombardia.

In particular we introduce a Bayesian model [Sect. 30.3] with a twofold scope: the evaluation of factors that are relevant in the evolution of the sequences of the hospitalizations process, and then the inference on inter-hospitalizations' times.

30.2 The Dataset

The dataset contains information about recurrent hospitalizations of patients affected by chronic heart disease. In particular, for each patient $j = 1, \ldots G$, we consider the sequence of hospitalizations and the times between two successive events. The analysis is carried out for $G = 26,618$ patients examined from 1/1/2006 to 31/12/2010. For each of them we know the number $n_j (i = 1, \ldots, n_j)$ of events A_i^j (hospitalization or death) occurred and the sequence of times between two next events (T_i^j is the time between A_{i-1}^j and A_i^j). We are also informed about covariates that may influence the occurrence of events, i.e., clinical covariates like status (0 healthy, 1 not), age, and sex (1 female, 0 male).

30.3 The Model

We assume patients to be conditionally independent. Further assumptions are the following:

h1: There is at least one hospitalization for each patient, i.e., $A_1^j = 1$.
h2: Any event after the first can be hospitalization or death, i.e., $A_i^j = i$ or $A_i^j = M$ with $i \in \{2, \ldots, n_j\}$.
h3: The occurrence of event A_i^j depends only on the last past event.
h4: The distribution of time T_i^j depends on all the previous waiting times and on the last event only, i.e., T_i^j is independent from $A_k^j \ \forall k < i - 1$.
h5: If a patient is still alive at the end of the analysis, we define $T_{n_j+1}^j$ as the time of the first event occurred after the end of the study; this is a right censored random variable.

Remark. For this analysis we considered only patients experiencing at most $N = 10$ hospitalizations, in order to guarantee that data would include enough information for all events, so that all parameters can be properly inferred.

Then we define the likelihood of the model, where $\rho = (\beta, \gamma, \lambda, a, q, p)$ is the vector of parameters:

$$\mathscr{L}(T, A | \rho) = \prod_{j=1}^{G} \mathscr{L}((T_1^j, A_1^j), (T_2^j, A_2^j), \ldots, (T_{n_j}^j, A_{n_j}^j), T_{n_j+1}^j | \rho)$$

and, for each patient $j \in \{1, \ldots, G\}$:

$$\begin{aligned}
\mathscr{L}(T^j, A^j | \rho) = &\mathscr{L}(T_1^j | \rho) \times \\
&\mathscr{L}(A_2^j | T_1^j, T_2^j, A_1^j = 1, \rho) \ \mathscr{L}(T_2^j | T_1^j, A_1^j = 1, \rho) \times \\
&\mathscr{L}(A_3^j | T_1^j, T_2^j, T_3^j, A_2^j, \rho) \ \mathscr{L}(T_3^j | T_1^j, T_2^j, A_2^j, \rho) \times \\
&\mathscr{L}(A_4^j | T_1^j, \ldots, T_4^j, A_3^j, \rho) \ \mathscr{L}(T_4^j | T_1^j, T_2^j, T_3^j, A_3^j, \rho) \times \\
&\ldots \ldots \times \\
&\mathscr{L}(A_{n_j}^j | T_1^j, T_2^j, \ldots, T_{n_j}^j, A_{n_j-1}^j, \rho) \times \\
&\mathscr{L}(T_{n_j}^j | T_1^j, \ldots, T_{n_j-1}^j, A_{n_j-1}^j, \rho) \times \\
&\mathscr{L}(T_{n_j+1}^j | T_1^j, \ldots, T_{n_j}^j, A_{n_j}^j, \rho).
\end{aligned}$$

Now, defining each term above, we have:

- $(T_1^j|\rho) \sim$ Weibull(λ_1, μ_1^j) being $\mu_1^j = \exp\{z_j^T \gamma\}$ and $z_j = (1, \text{age}_j, \text{sex}_j)$; $\gamma = (\gamma_0, \gamma_1, \gamma_2)$.
- $\mathscr{L}(A_2^j|T_1^j = t_1, T_2^j = t_2, A_1^j = 1, \rho) = \begin{cases} p_2 & \text{if } A_2^j = 2 \\ 1 - p_2 & \text{if } A_2^j = M. \end{cases}$
- $(T_2^j|T_1^j = t_1, A_1^j = 1, \rho) \sim$ Weibull(λ_2, μ_2^j) where $\mu_2^j = \exp\{\beta^T x_2^j\}$ and $x_2^j = (1, \text{age}_j, \text{sex}_j, \text{cov}_2^j, t_1)$, $\beta = (\beta_0, \beta_1, \beta_2, \beta_3, \beta_4)$. Note that we introduce the explicit dependence on the past here, through the parameter β_4.
- $\mathscr{L}(A_i^j|T_1^j = t_1, T_2^j = t_2, \ldots, T_i^j = t_i, A_{i-1}^j = i-1, \rho) = \begin{cases} p_i & \text{if } A_i^j = i \\ 1 - p_i & \text{if } A_i^j = M \end{cases}$
  $\forall i \in \{3, \ldots, n_j\}$. According to the meaning of each state A_i^j we specify only $A_{i-1}^j = i - 1$; if $A_{i-1}^j = M$ the event A_i^j will not occur. In fact for each patient j we know the total number of events n_j.
- $(T_i^j|T_1^j = t_1, \ldots, T_{i-1}^j = t_{i-1}, A_{i-1}^j = i - 1, \rho) \sim$ Weibull(λ_i, μ_i^j) where $\mu_i^j = \exp\{\beta^T x_i^j\}$, $x_i^j = (1, \text{age}_j, \text{sex}_j, \text{cov}_i^j, t_1 + \ldots + t_{i-1})$, $\beta = (\beta_0, \beta_1, \beta_2, \beta_3, \beta_4)$.
- $(T_{n_j+1}^j|T_1^j = t_1, \ldots, T_{n_j}^j = t_{n_j}, A_{n_j}^j = n_j, \rho) \sim \frac{\text{Weibull}(\lambda_{n_j}, \mu_{n_j}^j)}{S(\overline{T} - \sum_{i=1}^{n_j} t_i)} \times$
  $\mathbb{I}_{(\overline{T} - \sum_{i=1}^{n_j} t_i; +\infty)}\{T_{n_j+1}^j\}$ where $\mu_{n_j}^j = \exp\{\beta^T x_{n_j}^j\}$ and $x_{n_j}^j = (1, \text{age}_j, \text{sex}_j, \text{cov}_{n_j}^j, \sum_{k=1}^{n_j} t_k)$, $\beta = (\beta_0, \beta_1, \beta_2, \beta_3, \beta_4)$.

To complete the definition of the model we introduce the prior distribution of the parameter ρ; in particular we assume:

$$\pi(\rho) = \pi(p) \, \pi(\beta) \, \pi(\gamma) \, \pi(\lambda) \, \pi(a) \, \pi(q)$$

where $\pi(p) = \prod_{k=2}^{N} \pi(p_k)$; $p_i \sim$ Beta$(\alpha_i(t_1, \ldots, t_i) q_i(t_1, \ldots, t_i), \alpha_i(t_1, \ldots, t_i)$
$(1 - q_i(t_1, \ldots, t_i)))$ and then $\alpha_i = a/\{(t_1 + \ldots + t_{i-1}) t_i\}$ and $q_i(t_1, \ldots, t_i) = q_i$
$\pi(\beta) = \prod_{k=0}^{4} \pi(\beta_k)$ and marginally $\beta_k \sim N(0, 1000)$ $\forall k \in \{0, 1, \ldots, 4\}$
$\pi(\gamma) = \prod_{k=0}^{2} \pi(\gamma_k)$ and marginally $\gamma_k \sim N(0, 1000)$ $\forall k \in \{0, 1, 2\}$
$\pi(\lambda) = \prod_{k=1}^{N-2} \pi(\lambda_k)$ and marginally $\lambda_k \sim$ Gamma(η, ν) and $\eta = 10$, $\nu = 5$;
$\forall k \in \{1, 2, \ldots, N - 1\}$ and $\lambda_N = \lambda_{N-1} = \lambda_{N-2}$
$\pi(a) = $ Gamma$(2, 4)$
$\pi(q) = \prod_{k=2}^{N-2} \pi(q_k)$ and marginally $q_k \sim U(0, 1)$, $\forall k \in \{2, 3, \ldots, N - 1\}$
and $q_N = q_{N-1} = q_{N-2}$.

30.4 Posterior Inference

In this section we present the analysis of the posterior inference obtained with a Gibbs sampling algorithm run over 100,000 iterations with thinning of 10 iterations, discarding 1000 iterations as burn-in (model fitting was implemented in JAGS [2, 3]).

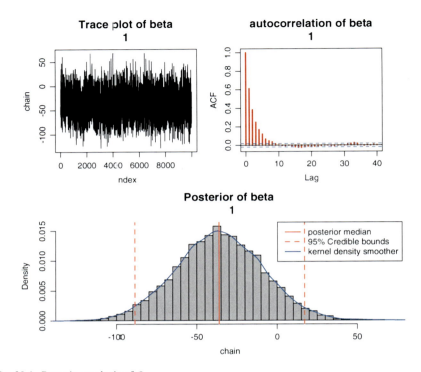

Fig. 30.1 Posterior analysis of β_1

Table 30.1 Posterior summaries of β

	Mean	SD	2.5%	50%	97.5%
β_0	−35.10	27.67	−89.91	−34.88	19.04
β_1	−1.15	0.40	−1.91	−1.15	−0.36
β_2	−21.82	13.06	−47.34	−21.85	3.50
β_3	−16.14	4.03	−27.06	−16.17	−8.27
β_4	67.55	7.18	53.70	67.53	81.87

The trace plots of parameters β, γ and λ confirm convergence, even starting from different initial points. In Fig. 30.1 we report trace, autocorrelation, and posterior distribution plots for β_1.

The shape parameters $\{\lambda_i\}$ of the Weibull distributions are not reported here, but they are concentrated around 1. On the other hand, Table 30.1 and Table 30.2 report the posterior summaries of the β and γ parameters; in particular we can see that women (sex=1) and unhealthy (clinical=0) patients have shorter times between hospitalizations (great evidence for $\beta_2 < 0$ and $\beta_3 < 0$) and that the longer is the time from first hospitalization to the last event, the longer will be the waiting time for the new event ($\beta_4 > 0$). After performing Bayesian model selection, we concluded that age can be excluded from the likelihood.

Table 30.2 Posterior summaries of γ

	Mean	SD	2.5%	50%	97.5%
γ_0	−15.62	27.97	−70.00	−15.45	39.89
γ_1	−1.08	0.42	−1.89	−1.08	−0.26
γ_2	−3.20	17.87	−38.06	−3.29	31.85

Considering the law of the sequence of events, instead, we observe that neither the chain of a nor the chains of $\{q_i\}$ converge well; further work is needed to better represent the conditional law of A_i^j, given ρ.

30.5 Conclusion

We believe that this model describes the evolution in time well enough, also taking into account the relevant covariates in the hospitalization process. As a further work, we could consider a different model, where, for example, we could define the $\{p_i\}$ in the likelihood as decreasing in i, ignoring the dependence on time.

References

1. Aalen O, Borgan O, Gjessing H (2008) Survival and event history analysis: a process point of view. Springer, New York
2. Lunn D, Jackson C, Best N, Thomas A, Spiegelhalter D (2012) The BUGS Book-A practical introduction to Bayesian analysis. CRC Press
3. Plummer M (2003) JAGS: a program for analysis of Bayesian graphical models using Gibbs sampling. In: Proceedings of the 3rd international workshop on distributed statistical computing, Vienna, Austria, 20–22 Mar 2003

Chapter 31
A Semiparametric Bayesian Multivariate Model for Survival Probabilities After Acute Myocardial Infarction

Elena Prandoni, Alessandra Guglielmi, Francesca Ieva, and Anna Maria Paganoni

Abstract In this work, a Bayesian semiparametric multivariate model is fitted to study data related to in-hospital and 60-day survival probabilities of patients admitted to a hospital with ST-elevation myocardial infarction diagnosis. We consider a hierarchical generalized linear model to predict survival probabilities and a process indicator (time of intervention). Poisson-Dirichlet process priors, generalizing the well-known Dirichlet process, are considered for modeling the random-effect distribution of the grouping factor which is the hospital of admission.

31.1 Introduction

The disease we are interested in is ST-elevation myocardial infarction (STEMI): it is caused by an occlusion of a coronary artery which causes an ischemia that, if untreated, can damage heart cells and make them die (infarction). It is very important that a reperfusion therapy could be done as quickly as possible, because its benefits decrease with delay in treatment; in our case, patients are treated with percutaneous transluminal coronary angioplasty. We consider data collected in the STEMI Archive [1], a multicenter observational prospective clinical study planned within the Strategic Program of Regione Lombardia. Data is recorded in a registry collecting clinical and process indicators, outcomes, and personal information on patients admitted to all hospitals of Regione Lombardia with STEMI diagnosis.

The regression model we introduce is multivariate, where the response has three components: door to balloon time (DB), i.e., the time between the admission to the hospital and angioplasty, on the logarithmic scale, in-hospital survival, and survival after 60 days from admission. The first term is an important indicator of

---

E. Prandoni (✉) • A. Guglielmi • F. Ieva • A.M. Paganoni
Politecnico di Milano, via Bonardi 9, 20133, Milano, Italy
e-mail: elenaprandoni@gmail.com

the efficiency of the health providers and plays a key role in the success of the therapy; the second one is the basic indicator of success or failure of the treatment, while the third one is a very important outcome, since doctors believe that it is in a 60-day period the effectiveness of the treatment in terms of survival and quality of life can be truly evaluated. We include the hospital random-effect parameters in the model and assume they are a sample from a Poisson-Dirichlet process a priori in order to eventually cluster the hospitals.

The main statistical aim of this work is prediction of both survival probabilities of new patients.

31.2 The Bayesian Model in a Nutshell

For each patient ($i = 1, \ldots, 697$) let $\mathbf{Y}_i := (Y_{i1}, Y_{i2}, Y_{i3})$ be the response, where Y_1 is the logarithm of DB, Y_2 is the in-hospital survival, and Y_3 is the long-term survival. We assume that observations, given parameters and covariates, are independent and the law of the response can be factorized into three parts:

$$\mathscr{L}(\mathbf{Y}_i|par, cov) = \mathscr{L}(Y_{i1}|par_1, cov_1)\mathscr{L}(Y_{i2}|Y_{i1}, par_2, cov_2)\mathscr{L}(Y_{i3}|Y_{i2}, par_3, cov_3).$$

The likelihood can be expressed as

$$Y_{i1}|\mu_i, \sigma \overset{ind}{\sim} \mathcal{N}(\mu_i, \sigma^2), \quad \mu_i = \sum_{l=1}^{4} \beta_l u_{il} + \beta_5 x_{i5} + \beta_6 x_{i6} \tag{1}$$

$$Y_{i2}|p_i, Y_{i1} \overset{ind}{\sim} Be(p_i), logit(p_i) = \alpha_1 z_{i1} + \alpha_2 z_{i2} + \alpha_3 Y_{i1} + \sum_{l=4}^{7} \alpha_l v_{il} + b_{\phi_{k[i]} k[i]} \tag{2}$$

$$Y_{i3}|r_i, Y_{i2} \overset{ind}{\sim} \begin{cases} Be(r_i) & \text{se } Y_{i2} = 1 \\ \delta_0 & \text{se } Y_{i2} = 0 \end{cases}, \quad logit(r_i) = \sum_{j=1}^{4} \gamma_j s_{ij} + t_{\phi_{k[i]} k[i]}. \tag{3}$$

Here $k[i]$ denotes the hospital where the patient i is admitted to, while covariates include the type of rescue unit sent to the patient ($u_{il}, l = 1, \ldots, 4$), the time of the first ECG (x_{i5}), the age (z_{i1}), the Killip class ($v_{il}, l = 1, \ldots, 4$, which quantify in four categories the severity of infarction), other covariates measuring the health status of the patient, or if the treatment was successful or not. The indexes $\phi_{k[i]}$ of the random-effect parameters in (2) and (3) assume values 1 or 0, if the hospital is in Milano or not.

As far as the fixed-effects parameters are concerned, we considered a parametric prior; for the random-effect parameters we assume Poisson-Dirichlet process priors with parameters (f, g) [2]. This choice helps us to avoid dependency on parametric assumptions and to increase flexibility in the prior and more robust inferences. We obtained posterior estimates of all the parameters through a Gibbs sampler algorithm implemented in JAGS [3] and with the support of R [4]; for this purpose we used a truncated stick-breaking representation of the nonparametric prior, which is a generalization of the well-known Dirichlet process [5].

Acknowledgements This work is within the Strategic Program "Exploitation, integration and study of current and future health databases in Lombardia for Acute Myocardial Infarction" supported by "Ministero del Lavoro, della Salute e delle Politiche Sociali" and by "Direzione Generale Sanità - Regione Lombardia." The authors wish to thank the Working Group for Cardiac Emergency in Milano, the Cardiology Society, and the 118 Dispatch Center.

References

1. Ieva F (2013) Designing and mining a multicenter observational clinical registry concerning patients with Acute Coronary Syndromes. In: Grieco, N, Marzegalli M, Paganoni AM (eds) New diagnostic, therapeutic and organizational strategies for acute coronary syndromes patients. Springer, Heidelberg
2. Pitman J, Yor M (1997) The two-parameter Poisson Dirichlet distribution derived from a stable subordinator. The Ann Probab 2:855–900
3. Plummer M (2003) JAGS: A program for analysis of Bayesian graphical models using Gibbs sampling. In Proceedings of the 3rd International Workshop on Distributed Statistical Computing, pp 20–22
4. R Development Core Team (2009) R: A Language and Environment for Statistical Computing. R Foundation for Statistical Computing, Vienna.
5. Sethuraman J (1994) A constructive definition of Dirichlet process prior. Statistica Sinica 2: 639–650

Chapter 32
Particle Learning Approach to Bayesian Model Selection: An Application from Neurology

Simon Taylor, Gareth Ridall, Chris Sherlock, and Paul Fearnhead

Abstract An improved method is sought to accurately quantify the number of motor units that operate a working muscle. Measurements of a muscle's contractive potential are obtained by administering a sequence of electrical stimuli. However, the firing patterns of the motor units are non-deterministic and therefore estimating their number is non-trivial. We consider a state-space model that assumes a *fixed* number of motor units to describe the hidden processes within the body. Particle learning methodology is applied to estimate the marginal likelihood for a range of models that assumes a different number of motor units. Simulation studies of these systems, containing up to 5 motor units, are very promising.

32.1 Introduction

We are interested in accurately quantifying the number of motor units (MUs) that supply a working muscle. A MU consists of a single motor neuron and the muscle fibres it governs. An electrical study of a muscle provides insight into the neuromuscular processes by measuring the compound muscle action potential (CAMP) for a range of stimuli. The ability to partition each CAMP into the contributions from each MU, a single motor unit potential (SMUP), is central to motor unit number estimation (MUNE). However, this is complicated by the occurrence of "alternation" [1]: where different MU combinations activate under identical conditions.

S. Taylor (✉) • G. Ridall • C. Sherlock • P. Fearnhead
Department of Mathematics and Statistics, Lancaster University, Lancaster, UK
e-mail: s.taylor2@lancaster.ac.uk; g.ridall@lancaster.ac.uk; c.sherlock@lancaster.ac.uk; p.fearnhead@lancaster.ac.uk

32.2 The Neuromuscular Model

We propose an adaptation to the state-space neuromuscular model [3] that describes the relationship between the applied stimulus, s_t for $t = 1, \ldots, T$, and the corresponding CAMP, y_t, through the biological processes. The state variable is defined to be the firing index vector, $\mathbf{k}_t = (k_{1,t}, \ldots, k_{j,t}, \ldots, k_{u,t})'$, where each element describes a single MU's reaction to the stimulus and the vector length, u, denotes the assumed *known* quantity of MUs within the system. The individual firing events are assumed to be independent Bernoulli random variables with a probability that depends on the administered stimulus via a non-decreasing link function, $F(\cdot\,;\cdot)$, with parameters specific to the MU, ϕ_j:

$$k_{j,t}|s_t, \phi_j \sim \text{Bernoulli}(\,F(s_t; \phi_j)\,) \,. \tag{1}$$

Each firing MU generates a SMUP that is assumed to be Gaussian with a unique mean, μ_j, but a common variance, σ^2. Denoting the mean vector of SMUPs as $\boldsymbol{\mu} = (\mu_1, \ldots, \mu_j, \ldots, \mu_u)'$, the recorded CAMP is the sum of the generated SMUPs plus a Gaussian baseline measure that has its own mean, μ_b, and variance, σ_b^2. We use the calibration data to approximate σ_b^2 and assume that $\sigma_b^2 \ll \sigma^2$. Hence, an indicator function, $I_{\{\cdot\}}$, may be introduced on the baseline event in defining the observation process:

$$y_t|\mathbf{k}_t, \mu_b, \sigma_b^2, \boldsymbol{\mu}, \sigma^2 \sim \text{N}\left(\,\mu_b + \mathbf{k}_t'\boldsymbol{\mu},\; \sigma_b^2 I_{\{\mathbf{k}_t=\mathbf{0}\}} + \sigma^2 \mathbf{1}'\mathbf{k}_t\,\right)\,. \tag{2}$$

32.3 Methodology

The neuromuscular model is used to conduct MUNE via Bayesian model selection, requiring reliable marginal likelihood estimates for a range of proposed models that assume a different number of MUs. Our approach to estimating the marginal likelihood, for a specific model, is to consider its predictive factorisation, where each term expresses the probability for a CAMP given the currently available data:

$$\Pr(y_{1:T}|s_{1:T}, u) = \Pr(y_1|s_1, u) \prod_{t=2}^{T} \Pr(y_t|y_{1:t-1}, s_{1:t}, u)\,. \tag{3}$$

Estimates of these terms are obtainable from independent applications of the particle learning methodology [2] to each considered model. This procedure is an extension of the auxiliary particle filter that constructs the particle set with the essential state vector (ESV), which contains the sufficient information necessary for the two-stage sequential procedure:

1. Resample the particles with weights proportional to the marginal predictive of y_t with all unknown parameters and state variables marginalised.

Table 32.1 Simulation study summaries from 100 hypothetical neuromuscular systems

Summary	Number of Motor Units (u)				
	1	2	3	4	5
$\hat{u} = u$ [a]	100%	100%	100%	100%	100%
95% Coverage[b]	100%	100%	100%	100%	100%
Mean Interval Width[c]	0.00	0.05	0.05	0.05	0.10

[a] Proportion of cases where the posterior mode estimate is the true value
[b] Proportion of cases where the "at least" 95% higher posterior density interval contains the true value
[c] Mean width of the "at least" 95% higher posterior density interval

2. Propagate the particles either deterministically or by generating appropriate random samples.

The marginal predictive terms are thereby estimated by Monte Carlo integration over the ESV within the procedure before the propagation stages.

32.4 Discussion

Our procedure has been applied to simulated data from 100 hypothetical neuromuscular systems that contain up to 5 MUs. The results, Table 32.1, illustrate that the posterior modal estimate correctly identified the true scenario for all data sets. The increase in the average interval width for larger systems illustrates that such systems are harder to analyse because there is a greater chance of incurring a period of alternation which involves multiple MUs, thus requiring more information to decipher the underlying structure.

Our aim is to adapt this procedure to analyse larger neuromuscular systems. However, the event space for $\mathbf{k}_t$ increases exponentially as larger models are considered. This substantially increases the computational complexity due to the need to marginalise all unknowns, parameters and states, within the algorithmic procedure.

References

1. Brown W, Milner-Brown H (1976) Some electrical properties of motor units and their effects on the methods of estimating motor unit numbers. J Neurol Neurosurg Psychiatry 39:249–257
2. Carvalho C, Johannes M, Lopes H (2010) Polson N Particle learning and smoothing. Statist Sci 25:88–106
3. Ridall P, Pettitt A, Henderson R, McCombe P (2006) Motor unit number estimation—a Bayesian approach. Biometrics 62: 1235–1250

Part V
Bayesian Models for Stochastic and Economic Processes

Chapter 33
Analysis of Italian Financial Market via Bayesian Dynamic Covariance Models

Daniele Durante

Abstract The attempt to provide a quantitative view on the evolution of the temporal and geo-economic relations between the Italian Stock Market Index FTSE MIB and the major financial markets before and during the global financial crisis of 2007–2012 motivates the search for statistical methodologies able to accommodate flexible dynamic structure of dependency among assets and to answer the main issues of multivariate financial time series analysis. This work compares, through an application study, some recent advances in Bayesian covariance regression, with a particular interest in the local adaptive smoothing of the stochastic processes under investigation in order to allow the covariances among returns to vary flexibly over continuous time.

33.1 Bayesian Covariance Regression for Financial Data

Spurred by the recent growth of interest in the dynamic dependence structure between financial markets in different countries and in its features during the crises that have followed in recent years, we focus our attention in the temporal evolution of the conditional correlations between the Italian Financial Market and those of the main countries involved during the recent crises.

To address this applicative problem we analyze the multivariate weekly time series of the main 33 National Stock Indices from 12/07/2004 to 25/06/2012, placing ourselves in the financial context where large datasets and high-frequency data motivate the search for a formulation able to handle high-dimensional data through tractable computations and simple online updating and prediction procedures. Besides these issues, it is important that the model allows for the presence

---

D. Durante (✉)
University of Padua, Department of Statistical Sciences, via Cesare Battisti,
241, 35121, Padua, Italy
e-mail: durante@stat.unipd.it

of missing values and takes also into account the possibility that covariances and variances can change rapidly during times of financial crisis, revealing different associations among assets and countries than occur in a healthier economic climate.

There is a rich literature on univariate stochastic volatility modeling, with an increasing emphasis on multivariate generalizations. Two recent answers from a Bayesian perspective are provided by the Bayesian Nonparametric Covariance Regression (BCR) model [3] and the Locally Adaptive Bayesian Covariance Regression (LBCR) [2]. Both approaches define the covariance matrix of a vector of p variables at time t_i, as a regularized quadratic function of time-varying loadings in a latent factor model, characterizing the latter as a sparse combination of a collection of unknown dictionary functions. More specifically given a set of $p \times 1$ vectors of observations $y_i \sim N_p(\mu(t_i), \Sigma(t_i))$ where $i = 1, \ldots, T$ indexes time, both models define

$$\operatorname{cov}(y_i | t_i = t) = \Sigma(t) = \Theta \xi(t) \xi(t)^T \Theta^T + \Sigma_0, \quad t \in \mathscr{T} \subset \Re^+, \tag{1}$$

where Θ is a $p \times L$ matrix of coefficients, $\xi(t)$ is a time-varying $L \times K$ matrix with unknown continuous dictionary functions entries $\xi_{lk} : \mathscr{T} \to \Re$, and finally Σ_0 is a positive definite diagonal matrix. The previous equation for $\Sigma(t)$ results from the marginalization of η_i in the latent factor model

$$y_i = \Theta \xi(t_i) \eta_i + \epsilon_i, \tag{2}$$

with the latent factors $\eta_i \sim N_K(0, I_K)$ and $\epsilon_i \sim N_p(0, \Sigma_0)$. A further generalization of the model allows also for the possibility of including the nonparametric mean regression by assuming

$$\eta_i = \psi(t_i) + \nu_i, \tag{3}$$

where $\nu_i \sim N_K(0, I_K)$ and $\psi(t)$ is a $K \times 1$ matrix with unknown continuous entries $\psi_k : \mathscr{T} \to \Re$ that can be modeled in a related manner to the dictionary elements in $\xi(t)$. The induced mean of y_i conditionally on $t_i = t$ and marginalizing out ν_i is then

$$\mu(t) = \Theta \xi(t) \psi(t). \tag{4}$$

Although the two approaches are the same in terms of model formulation and specification of the priors Π_Θ, Π_{Σ_0} for the parametric components of the model Θ and Σ_0, they differ substantially with respect to the choice of the priors Π_ξ and Π_ψ for $\xi_{\mathscr{T}} = \{\xi(t), t \in \mathscr{T}\}$, and $\psi_{\mathscr{T}} = \{\psi(t), t \in \mathscr{T}\}$, respectively. Fox and Dunson [3] consider the dictionary functions ξ_{lk} and ψ_k as independent Gaussian processes (GP) $GP(0, c)$ with c as the squared exponential correlation function having $c(x, x') = \exp(-k||x-x'||_2^2)$. This approach provides a continuous time and flexible model that accommodates missing data and scales to large p, but the proposed prior for the dictionary functions assumes a stationary dependence structure and hence induces priors distributions Π_Σ and Π_μ on $\Sigma_\mathscr{T}$ and $\mu_\mathscr{T}$ through (1) and (4) that tend to under-smooth during periods of stability and over-smooth during periods of sharp change.

To answer this issue, Durante, Scarpa, and Dunson [2] develop a covariance stochastic process with locally varying smoothness by replacing GP prior for $\xi_\mathscr{T} = \{\xi(t), t \in \mathscr{T}\}$ and $\psi_\mathscr{T} = \{\psi(t), t \in \mathscr{T}\}$ with nested Gaussian process (nGP) priors [4], with the goal of maintaining simple computation and allowing both covariances and means to vary flexibly over continuous time. The nGP provides a highly flexible prior on the dictionary functions whose smoothness, explicitly modeled by their derivatives via stochastic differential equations, is expected to be centered on a local instantaneous mean function, which represents a higher-level GP, that induces adaptivity to locally varying smoothing.

Restricting our attention on Π_ξ (the same holds for Π_ψ), the Markovian property implied by the stochastic differential equations allows a simple state space formulation of nGP in which the prior for $\xi_{lk}(t)$ along with its first-order derivative $\xi'_{lk}(t)$ and the locally instantaneous mean $A_{lk}(t) = \mathrm{E}[\xi'_{lk}(t)|A_{lk}(t)]$ follow the approximated state equation:

$$\begin{bmatrix} \xi_{lk}(t_{i+1}) \\ \xi'_{lk}(t_{i+1}) \\ A_{lk}(t_{i+1}) \end{bmatrix} = \begin{bmatrix} 1 & \delta_i & 0 \\ 0 & 1 & \delta_i \\ 0 & 0 & 1 \end{bmatrix} \begin{bmatrix} \xi_{lk}(t_i) \\ \xi'_{lk}(t_i) \\ A_{lk}(t_i) \end{bmatrix} + \begin{bmatrix} 0 & 0 \\ 1 & 0 \\ 0 & 1 \end{bmatrix} \begin{bmatrix} \omega_{i,\xi_{lk}} \\ \omega_{i,A_{lk}} \end{bmatrix}, \quad (5)$$

where $[\omega_{i,\xi_{lk}}, \omega_{i,A_{lk}}]^T \sim N_2(0, V_{i,lk})$, with $V_{i,lk} = \mathrm{diag}(\sigma^2_{\xi_{lk}}\delta_i, \sigma^2_{A_{lk}}\delta_i)$ and $\delta_i = t_{i+1} - t_i$. This formulation allows the implementation of an online updating algorithm and facilitates the definition of a simple Gibbs sampling for posterior computation which alternates between a simulation smoother step [1] to update the nGP prior and standard Gibbs sampling steps for the parametric component of the model.

33.2 Application to National Stock Indices (NSI)

To compare LBCR and BCR in applicative settings we consider, for both of them, the heteroscedastic model for the log returns $y_i \sim N_{33}(\mu(t_i), \Sigma(t_i))$ with $i = 1, \ldots, 415$ and t_i in the discrete set $\mathscr{T}_o = \{1, 2, \ldots, 415\}$, where mean $\mu(t_i)$ and covariance matrix $\Sigma(t_i)$ of the National Stock Indices at time $t = t_i$ are given in (4) and (1), respectively. We run 10,000 Gibbs iterations with a burn-in of 2,500, choosing diffuse but proper priors.

The time-varying estimated correlations with respect to Italy FTSE MIB in Fig. 33.1 show the presence of an evident geo-economic structure in world markets, which is more evident in LBCR than in BCR. Note also that the latter, as expected, tends to over-smooth the dynamic dependence structure during the financial crisis, proving to be not able to model the dramatic change in the correlations between

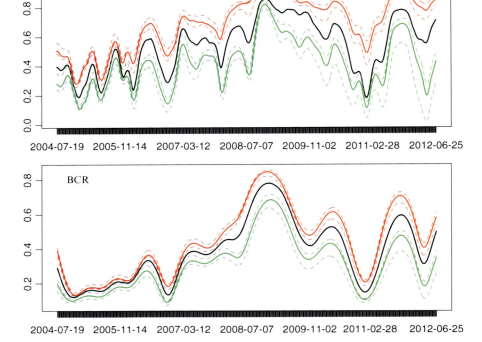

Fig. 33.1 *Black line*: For ITA FTSE MIB median of correlations with the other 32 NSI based on posterior mean of $\{\Sigma(t_i)\}_{i=1}^{415}$. Red lines: 25%, 75% (*dotted lines*), and 50% (*solid line*) quantiles of correlations between ITA FTSE MIB and European countries (we included also the USA). Green lines: 25%, 75% (*dotted lines*), and 50% (*solid line*) quantiles of correlations between ITA FTSE MIB and the Asian Tigers

Italy and Economic Tigers during the late 2008, and the two peaks representing, respectively, Irish and Portugal debt crisis at the beginning of 2011. In general, correlations among financial markets increase significantly during the crises, showing a clear international financial contagion effect in agreement with other theories on financial crises. LBCR highlights the persistence of high levels of correlation during the global financial crisis between the late 2008 and end 2009, with further rapid changes in correspondence of Greek crisis at the beginning of 2010, and in mid-2011 with the worsening of European sovereign-debt crisis and the rejection of US budget.

References

1. Durbin J, Koopman S (2002) A simple and efficient simulation smoother for state space time series analysis. Biometrika 89:603–616
2. Durante D, Scarpa B, Dunson DB (2012) Locally adaptive Bayesian covariance regression. Available via arXiv. http://arxiv.org/abs/1210.2022v1
3. Fox E, Dunson DB (2011) Bayesian Nonparametrics Covariance Regression. Available via arXiv. http://arxiv.org/abs/1101.2017
4. Zhu B, Dunson DB (2012) Locally Adaptive Bayes Nonparametric Regression via Nested Gaussian Processes. Available via arXiv. http://arxiv.org/abs/1201.4403

Chapter 34
Bayesian Model Selection of Regular Vine Copulas

Lutz F. Gruber and Claudia Czado

Abstract Regular vine copulas can describe a wider array of dependency patterns than the multivariate Gaussian copula or the multivariate Student's t copula. We present reversible jump Markov chain Monte Carlo algorithms to estimate the joint posterior distribution of the density factorization, pair copula families, and parameters of a regular vine copula. A simulation study shows that our algorithms outperform model selection methods suggested in the current literature and succeed in selecting the true model when other methods fail. Furthermore, we present an application study that shows how a vine copula-based approach can improve the pricing of exotic financial derivatives.

34.1 Introduction

Multivariate data with rich patterns of dependence are found in many fields in business and science. Copulas, the tool of choice to model these patterns of dependence, are multivariate distributions with uniform margins [9]. The Gaussian copula, possibly the most widely known copula, even appears in the mainstream media as "The Formula That Killed Wall Street" [8]. The signature feature of copulas is that they allow dependence characteristics to be modeled separately from the marginal distributions. This provides the added benefit that copulas can be introduced to existing models that do not yet incorporate measures of dependence, but feature established models for the margins.

L.F. Gruber (✉) • C. Czado
Technische Universität München, Zentrum Mathematik, Lehrstuhl für Mathematische Statistik, München, Germany
e-mail: lutz.gruber@ma.tum.de; cczado@ma.tum.de

34.2 Regular Vine Copulas

While many classes of bivariate copulas, also called pair copulas, are well known [6], there are only a limited number of multivariate copulas available with a closed-form analytical expression. However, these multivariate copulas cover only limited patterns of dependence. As a pair copula construction with arbitrary pair copulas, regular vine copulas can describe a much wider array of multivariate dependencies.

Specifically, regular vine copulas are set up in two steps. First is the construction of an n-dimensional copula density from (conditional) pair copula densities. These pair copulas are organized in a sequence of linked trees $\mathscr{V} = (T_1, \ldots, T_{n-1})$, which is called the regular vine. Each of the $n - j$ edges of tree T_j, $1 \leq j \leq (n-1)$, corresponds to a bivariate copula density that is conditional on $j - 1$ variables. Secondly, a copula family is selected for each of these (conditional) bivariate building blocks from a set of bivariate (parametric) candidate families **B**. We denote the mapping of the pair copulas to the regular vine by $\mathscr{B}_\mathscr{V}(\boldsymbol{\theta}_{\mathscr{B}_\mathscr{V}})$, where we write $\boldsymbol{\theta}_{\mathscr{B}_\mathscr{V}}$ for the parameters of the pair copulas.

34.3 Model Selection Algorithm

We present a reversible jump MCMC sampler [2] for Bayesian model selection of regular vine copulas. Our priors enforce model sparsity, but do not make structural assumptions about the vine copula. More specifically, we assume

$$\mathscr{V} \sim discrete\, Uniform(\cdot),$$
$$\mathscr{B}_\mathscr{V} \mid \mathscr{V} \sim \exp(-\lambda d_{\mathscr{B}_\mathscr{V}}), \lambda \geq 0,$$
$$\boldsymbol{\theta}_{\mathscr{B}_\mathscr{V}} \mid \mathscr{V}, \mathscr{B}_\mathscr{V} \sim Uniform(\cdot),$$

where $d_{\mathscr{B}_\mathscr{V}}$ denotes the number of parameters of the regular vine copula $\mathscr{V}(\mathscr{B}_\mathscr{V}(\boldsymbol{\theta}_{\mathscr{B}_\mathscr{V}}))$ or, equivalently, the dimension of the parameter vector $\boldsymbol{\theta}_{\mathscr{B}_\mathscr{V}}$. The prior density $\pi(\mathscr{V}(\mathscr{B}_\mathscr{V}(\boldsymbol{\theta}_{\mathscr{B}_\mathscr{V}})))$ of a regular vine copula then follows

$$\pi(\mathscr{V}(\mathscr{B}_\mathscr{V}(\boldsymbol{\theta}_{\mathscr{B}_\mathscr{V}}))) \propto \exp(-\lambda d_{\mathscr{B}_\mathscr{V}}).$$

Our algorithm to sample from the posterior distribution of the regular vine copula $\mathscr{V}(\mathscr{B}_\mathscr{V}(\boldsymbol{\theta}_{\mathscr{B}_\mathscr{V}}))$ given observed data follows the general outline below [3]. Furthermore, we developed a tree-by-tree version of this model selection algorithm that allows for faster computation times [4]. The tree-by-tree approach achieves a significant reduction in computational complexity by reducing the model search space by many orders of magnitude.

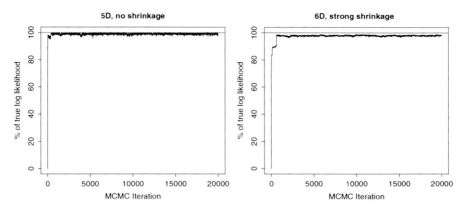

Fig. 34.1 Log likelihood trace plots of a simulation study with known 5-dimensional and 6-dimensional regular vine copulas

General Reversible Jump MCMC-Based Model Selection

1: Choose an arbitrary regular vine copula as the starting value.
2: **for each** MCMC iteration $r = 1, \ldots, R$ **do**
3:   Perform a within-model move: perform a Metropolis–Hastings update [5, 7] of the parameters $\theta_{\mathcal{B}_\mathcal{V}}$.
4:   Perform a between-models move: update the regular vine $\mathcal{V}$ along with, or only, the pair copulas $\mathcal{B}_\mathcal{V}(\theta_{\mathcal{B}_\mathcal{V}})$.
5: **end for**
6: **return** the Bayesian posterior sample $\left(\mathcal{V}^r(\mathcal{B}_\mathcal{V}^r(\theta_{\mathcal{B}_\mathcal{V}}^r))\right)_{r=1,\ldots,R}$.

34.4 Simulation Study and Application Study

The proposal mechanisms of our sampling algorithms are designed to achieve rapid convergence of the MCMC sampling chain (Fig. 34.1). Depending on the strength of the shrinkage prior, our algorithm recovers 98% to 100% of the log likelihood of the "true" models used in our simulation studies. This represents a major improvement over existing model selection methods whose recovery rates are in the 75% to 85% range.

Additionally, we estimated the expected payout of an exotic option on a basket of nine securities. First, we strip the return time series of their time dependencies using GARCH(1,1) models [1]. Then we estimate a regular vine copula to the approximately independent *Uniform(0,1)* data using our reduced tree-by-tree model selection algorithm. We calculate Monte Carlo estimates of the payouts by drawing from the estimated copula's distribution for different barrier prices of the option (Fig. 34.2). Current best practice is to model dependencies in financial data with the

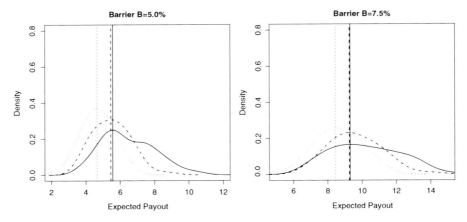

Fig. 34.2 Density estimates of the option payout with density modes. The *solid line* shows the bootstrapped observed data; *the dashed* and *dotted lines* correspond to the estimated vine copula and Student's t copula, respectively

Student's t copula. Our comparison of the payout estimates demonstrates that vine copulas are superior dependence models for financial data.

Reference

1. Fan J, Yao Q (2001) Nonlinear time series: nonparametric and parametric methods. Springer, New York
2. Green PJ (1995) Reversible jump Markov chain Monte Carlo computation and Bayesian model determination. Biometrika, 82:711–732
3. Gruber LF, Czado C (2013) Bayesian model selection of regular vine copulas. Working paper
4. Gruber LF, Czado C (2013) Sequential bayesian model selection of regular vine copulas. Submitted to Bayesian Analysis
5. Hastings, WK (1970) Monte Carlo sampling methods using Markov chains and their applications. Biometrika, 57:97–109
6. Joe H (2001) Multivariate models and dependence concepts. Chapman & Hall, London
7. Metropolis N, Rosenbluth AW, Rosenbluth MN, Teller AH, Teller E(1953) Equation of state calculations by fast computing machines. J Chemical Physics 21:1087–1092
8. Salmon F (2009) Recipe for disaster: the formula that killed wall street. Wired Magazine 17(3):74–79, 112
9. Sklar A (1959) Fonctions de répartition à n dimensions et leurs marges. Publications de l'Institut de Statistique de l'Université de Paris 8:229–231

Chapter 35
Analysis of Exchange Rates via Multivariate Bayesian Factor Stochastic Volatility Models

Gregor Kastner, Sylvia Frühwirth-Schnatter, and Hedibert F. Lopes

Abstract Multivariate factor stochastic volatility (SV) models are increasingly used for the analysis of multivariate financial and economic time series because they can capture the volatility dynamics by a small number of latent factors. The main advantage of such a model is its parsimony, as the variances and covariances of a time series vector are governed by a low-dimensional common factor with the components following independent SV models. For high-dimensional problems of this kind, Bayesian MCMC estimation is a very efficient estimation method; however, it is associated with a considerable computational burden when the dimensionality of the data is moderate to large. To overcome this, we avoid the usual forward-filtering backward-sampling (FFBS) algorithm by sampling "all without a loop" (AWOL), consider various reparameterizations such as (partial) noncentering, and apply an ancillarity-sufficiency interweaving strategy (ASIS) for boosting MCMC estimation at a univariate level, which can be applied directly to heteroskedasticity estimation for latent variables such as factors. To show the effectiveness of our approach, we apply the model to a vector of daily exchange rate data.

G. Kastner (✉) • S. Frühwirth-Schnatter
WU Vienna University of Economics and Business, Institute for Statistics and Mathematics, Welthandelsplatz 1, 1020 Wien, Austria
e-mail: gregor.kastner@wu.ac.at; sylvia.fruehwirth-schnatter@wu.ac.at

H.F. Lopes
The University of Chicago, Booth School of Business, 5807 South Woodlawn Avenue, Chicago IL 60637, USA
e-mail: hlopes@chicagobooth.edu

35.1 Introduction

In the recent years, factor SV models have been progressively applied to important problems in financial econometrics such as asset allocation and asset pricing. These models extend standard factor pricing models such as the arbitrage pricing theory and the capital asset pricing model. As opposed to factor SV models, standard factor pricing models do not attempt to model the dynamics of the volatilities of the asset returns and usually assume that the covariance matrix $\Sigma_t \equiv \Sigma$ is constant. Empirical evidence suggests that multivariate factor SV models are a promising approach for modeling multivariate time-varying volatility, explaining excess asset returns, and generating optimal portfolio strategies. Following [1], the model reads

$$y_t = \Lambda f_t + \Sigma_t^{1/2} \epsilon_t, \qquad \epsilon_t \sim N_m(0, I_m), \qquad (1)$$

$$f_t = V_t^{1/2} u_t, \qquad u_t \sim N_r(0, I_r), \qquad (2)$$

where for $t = 1, \ldots, T$, the vector $y_t = (y_{1t}, \ldots, y_{mt})'$ consists of (potentially demeaned) log returns of m observed time series, $\Sigma_t = \text{Diag}(\exp(h_{1t}), \ldots, \exp(h_{mt}))$, $V_t = \text{Diag}(\exp(h_{m+1,t}), \ldots, \exp(h_{m+r,t}))$, and Λ is an unknown $m \times r$ factor loading matrix with elements Λ_{ij}. The standard assumption is that f_t, f_s, ϵ_t, and ϵ_s are pairwise independent for all t and s. Both the latent factors and the idiosyncratic shocks are allowed to follow different stochastic volatility processes, i.e.,

$$h_{it} = \mu_i + \phi_i(h_{i,t-1} - \mu_i) + \sigma_i \eta_{it}, \qquad \eta_{it} \sim N(0, 1). \qquad (3)$$

In the following, we identify the model by imposing a lower-triangular structure for Λ with unconstrained diagonal elements and therefore set $\mu_i = 0$ for $i \in \{m+1, \ldots, m+r\}$. Clearly, this introduces an order dependence among the responses and makes the appropriate choice of the first r variables an important modeling decision.

35.2 Factor SV Estimation

After fixing $T(m + 2r) + mr + 4m + 3r$, in our application 81763, starting values for (the elements of) $\mu, \phi, \sigma, h, f, \Lambda$, we repeat the following steps:

(a) *Perform $m + r$ univariate SV updates* for $h_{i0}, \ldots, h_{iT}$, ϕ_i, σ_i and m updates for μ_i. We do this by sampling the latent variables AWOL as in [4]; thus, no FFBS methods are required, there is no need to invert the tridiagonal information matrix of the joint conditional distribution of the latent log volatilities and computations are fast due to the availability of band back-substitution already implemented in practically all widely used programming languages. Moreover,

we employ several variants of ASIS [5] by moving the parameters of interest from the state equation (3) in its centered parameterization to the augmented observation equation (1) or (2) and perform an extra update for these parameters in the noncentered parameterization. Doing so is very cheap in terms of computation—only around 2 % extra CPU time is needed—and nevertheless has substantial effect on sampling efficiency. The actual sampler is written in C and made accessible through the R package stochvol [2], publicly available on CRAN. More details about efficient univariate SV estimation can be found in [3].
(b) *Sample the factor loadings*, constituting m independent r-variate regression problems, from the T-dimensional Gaussian distribution $\Lambda_{i\cdot}|f, y_i, h_i$.
(c) *Sample the latent factors*, constituting T independent r-variate regression problems, from the m-dimensional Gaussian distribution $f_t|\Lambda, y_{\cdot t}, h_{\cdot t}$.

35.3 Application

We apply a three-factor SV model to EUR exchange rates, quoted indirectly. The data stems from the European Bank's Statistical Data Warehouse and comprises $T = 3140$ observations of 20 currencies ranging from January 3, 2000, to April 4, 2012. Figure 35.1 depicts the proportions of the variance which can be

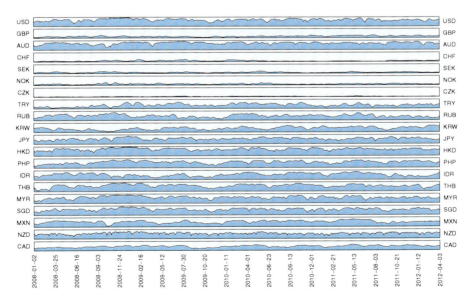

Fig. 35.1 Median posterior proportions of variances explained by the common latent factors given through $1 - \Sigma_{ii,t}/\mathrm{var}(y_{it})$. Results are displayed on a daily basis, for the time from 2008 onwards

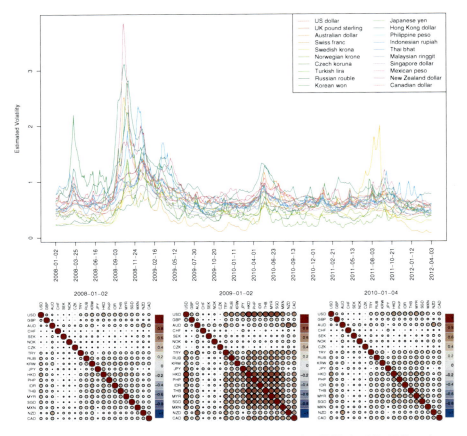

Fig. 35.2 Estimated median univariate volatilities in percent (after 2007) for daily EUR exchange rates (*top*) and median posterior correlation matrices as implied by $\text{cov}(\boldsymbol{y}_t) = \boldsymbol{\Lambda} \boldsymbol{V}_t \boldsymbol{\Lambda}' + \boldsymbol{\Sigma}_t$, exemplified for the first trading days in 2008, 2009, and 2010 (*bottom*)

explained through the common factors over the time period 2008–2012. Note that the explanatory power of the common factors varies strongly over currencies as well as time. In the top panel of Fig. 35.2, the individual latent volatilities are displayed for the same time period. They exhibit pronounced heteroskedasticity as well as considerable co-movement, providing more empirical evidence for multivariate modeling through common latent factors. The bottom panel of Fig. 35.2 features three examples of instantaneous correlation matrices. It stands out that practically all correlations are positive but again substantially time varying. Moreover, some clusters of highly correlated currencies (such as the "Asian Tigers") can be spotted, while continental European currencies show little correlation.

References

1. Chib S, Nardari F, Shephard N (2006) Analysis of high dimensional multivariate stochastic volatility models. J. Econom 134:341–371
2. Kastner G (2013) stochvol: Efficient Bayesian inference for stochastic volatility (SV) models. R package version 0.6-1
3. Kastner G, Frühwirth-Schnatter S (forthcoming) Ancillarity-sufficiency interweaving strategy (ASIS) for boosting MCMC estimation of stochastic volatility models. Comput Stat Data An doi: 10.1016/j.csda.2013.01.002
4. McCausland WJ, Miller S, Pelletier D (2011) Simulation smoothing for state-space models: a computational efficiency analysis. Comput Stat Data An 55:199–212
5. Yu Y, Meng X-L (2011) To center or not to center: that is not the question—an ancillarity-sufficiency interweaving strategy (ASIS) for boosting MCMC efficiency. J Comput Graph Stat 20:531–570

Chapter 36
On Some Stationary Models: Construction and Estimation

Consuelo R. Nava, Ramsés H. Mena, and Igor Prünster

Abstract We propose a simple yet powerful method to construct strictly stationary Markovian models with given, but arbitrary, invariant distributions. The idea is based on a Poisson transform modulating the dependence structure in the model. An appealing feature of our approach is that we are able to fully control the underlying transition probabilities and therefore incorporate them within standard estimation methods. We analyze some specific cases in both discrete and continuous time. Given our proposed representation of the transition density, a Gibbs sampler algorithm, based on the slice method, is proposed and implemented. In particular, the resulting methodology is of interest for the estimation of certain continuous time models, such as diffusion processes.

36.1 Markovian Models with Given Marginal Distributions

Stationary processes represent a key component in several modeling approaches used in probability and statistics, mainly because estimation and prediction procedures are more accessible than those for not stationary models. It is worth noting that by appropriately relaxing the distributional assumptions of the underlying stationary and transition distributions, then data features typically associated to nonstationary sequences can also be captured through stationary models, e.g., think of a Markovian model with bivariate stationary distribution. When considering random phenomena evolving in time a natural starting point is to consider

---

C.R. Nava (✉) • I. Prünster
Università degli Studi di Torino, Collegio Carlo Alberto, Turin, Italy
e-mail: consuelo.nava@carloalberto.com; igor.prunster@unito.it

R.H. Mena
Universidad Nacional Autónoma de México, Mexico city, Mexico
e-mail: ramses@sigma.iimas.unam.mx

Markovian processes and thus look for transition mechanisms that retain a particular distribution invariant over time. This is the approach followed by many of the constructions available in the literature, in both discrete and continuous time. Most of these approaches start from a stochastic equation describing the dynamics in time. Unfortunately, the availability of analytic expressions for the corresponding transition probabilities is not always immediate. However, a full control of the transition probabilities driving a Markovian process is desirable, especially for the conveyed advantages in estimation and prediction procedures. In [10] the reversibility property characterizing Gibbs sampler Markov chains is exploited to propose strictly stationary AR(1)-type models with virtually any choice of marginal distribution. In particular, they demonstrate that various general approaches, such as the one by [2], can be seen as a particular case. Being such a general approach, concrete choices of dependence should be made to accommodate specific modeling needs. Indeed, examples of this construction, meeting some specific dependency or distributional features, can be found in [1, 4, 6–8]. Of particular interest here is the approach to continuous time Markovian models studied in [9]. The idea presented in [3], and summarized in this note, aims at constructing stationary Markovian models using a Poisson transform. The resulting specific dependence structure is general enough. In particular, it allows to construct models with invariant distributions on $\mathbb{R}_+$, which can also be extended to processes supported on other state spaces.

36.2 Poisson Weighted Density

Let f be a continuous density function supported on $\mathbb{R}_+$. For any $y \in \{0, 1, 2, \ldots\}$ and $\phi > 0$, we define the *Poisson weighted density* as

$$\hat{f}(x; y, \phi) := \frac{x^y e^{-x\phi} f(x)}{\xi(y, \phi)} \quad \text{where} \quad \xi(y, \phi) := \int_{\mathbb{R}_+} z^y e^{-z\phi} f(z) \mathrm{d}z \qquad (1)$$

Notice that (1) constitutes a well-defined probability density on $\mathbb{R}_+$, provided the above integral exists. For $\phi \downarrow 0$, it reduces to the size-biased density of f and, when $y = 0$, it reduces to the Esscher transform of f. To construct a stationary Markovian process, $(X_n)_{n \in \mathbb{Z}_+}$, with invariant distribution having density f, we use the Poisson weighted density, defining the time-homogeneous one-step ahead Markovian density as follows:

$$p(x_{n-1}, x_n) = \exp\{-\phi(x_n + x_{n-1})\} f(x_n) \sum_{y=0}^{\infty} \frac{(x_n x_{n-1} \phi)^y}{y! \xi(y, \phi)} \qquad (2)$$

which clearly satisfies $p(x_{n-1}, x_n) f(x_{n-1}) = p(x_n, x_{n-1}) f(x_n)$ for all $x_{n-1}, x_n \in \mathbb{R}_+$, leading to a time-reversible Markovian process.

Definition 1. We term the stationary Markovian process, driven by transition density (2) and stationary density f, an f-stationary Poisson-driven Markov process.

Cases of interest, including constructions of diffusion models with gamma, generalized inverse gaussian, and generalized extreme value (GEV) marginals, are presented in [3].

36.2.1 Estimation

The availability of a tractable expression for the transition density is appealing in the analysis and estimation of Markov processes. In particular if the choice of f leads to a manageable analytic expression in (2), the maximum likelihood estimator (MLE) can be easily determined. Alternatively, one could make use of such a representation for the transition density and obtain an MLE via the expectation-maximization (EM) algorithm based on the augmented likelihood or a Gibbs sampler algorithm for Bayesian estimation. In particular, using some slicing ideas from [5], we propose a Gibbs sampler algorithm, under the assumption of a continuous time process, with the following representation of the augmented transition density:

$$p_t(x_0, x_t, u, y) = \mathbb{I}(u \leq \psi_y) \exp\{-\phi_t(x_t + x_0)\} f(x_t; \boldsymbol{\theta}) \frac{(x_t\, x_0\, \phi_t)^y}{y!\, \psi_y\, \xi(y, \phi_t; \boldsymbol{\theta})} \quad (3)$$

where $y \mapsto \psi_y$ is an $\mathbb{N}$-valued function with known inverse ψ^*, e.g., $e^{-\eta y}$, for $0 \leq \eta \leq 1$. The use of these latent variables allows us to construct the augmented likelihood for a set of observations $\mathbf{x} = (x_1, \ldots, x_N)$ at times $(t_1, \ldots, t_N)$. When the time function ϕ does not depend on the parameters in the stationary distribution, we could separate the parameters of interest as $\boldsymbol{\theta} = (\boldsymbol{\theta}^{(s)}, \boldsymbol{\theta}^{(t)})$. Therefore, together with the assumption of independent priors, e.g., $\pi(\boldsymbol{\theta}^{(s)})$ and $\pi(\boldsymbol{\theta}^{(t)})$, the full conditionals can be obtained from following the log-posterior distributions:

$$\log \pi(\boldsymbol{\theta}^{(s)} \mid \cdots) \propto \log \pi(\boldsymbol{\theta}^{(s)}) + \sum_{n=1}^{N} \log(f(x_n; \boldsymbol{\theta}^{(s)})) - \sum_{n=2}^{N} \log(\xi(y_n, \phi_{\tau_n}; \boldsymbol{\theta}))$$

$$\log \pi(\boldsymbol{\theta}^{(t)} \mid \cdots) \propto \log \pi(\boldsymbol{\theta}^{(t)}) + \sum_{n=2}^{N} y_n \log(\phi_{\tau_n}) - \sum_{n=2}^{N} \phi_{\tau_n}(x_n + x_{n-1})$$

$$- \sum_{n=2}^{N} \log(\xi(y_n, \phi_{\tau_n}; \boldsymbol{\theta})),$$

where $\tau_n := t_n - t_{n-1}$, $\mathbf{u} = (u_2, \ldots, u_N)$ and $\mathbf{y} = (y_2, \ldots, y_N)$. Simulation from these posterior distributions can be done via the adaptive rejection Metropolis

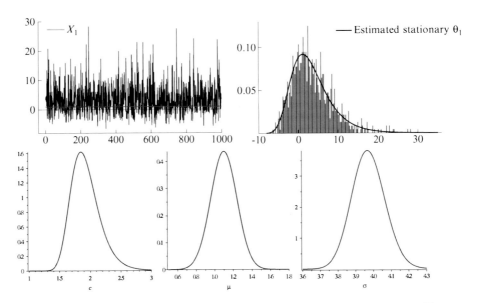

Fig. 36.1 Simulated data from a GEV I-stationary Poisson-driven Markov process. The *top plots* show the simulated time series (*left*) and the data histogram together with stationary GEV distribution (*right*). The *bottom plots* show the posterior estimates for $\theta_1 = (c, \mu, \sigma)$, with modes located at $(1.86, 1.09, 3.964)$, respectively

sampling (ARMS) algorithm. Furthermore, the full conditional distributions for the latent variables can be obtained via

$$\pi(u_n \mid \cdots) = \mathsf{U}(u_n; 0, \psi_{y_n})$$

$$\pi(y_n \mid \cdots) \propto \frac{[x_n \, x_{n-1} \, \phi_{\tau_n}]^{y_n}}{y_n! \, \xi(y_n, \phi_{\tau_n}; \boldsymbol{\theta}) \, \psi_{y_n}} \mathbb{I}(u_n \leq \psi_{y_n})$$

for $n = 2, \ldots, N$. Notice that the above distribution has support $y_n = 0, \ldots, \lfloor \psi^*(u_n) \rfloor$. This is the advantage of the slice method, i.e., that we only need to sample from a finite distribution. Figure 36.1 shows the posterior estimates of the above MCMC scheme for a GEV I-stationary Poisson-driven Markov process. The data were generated from the model with parameters $(c, \mu, \sigma) = (2, 1, 4)$. See [3].

References

1. Contreras-Cristán A, Mena RH, Walker SG (2009) On the construction of stationary AR(1) models via random distributions. Statistics 43:227–240

2. Joe H (1996) Time series models with univariate margins in the convolution-closed infinitely divisible class. J Appl Probab 33:664–677
3. Nava C, Mena RH, Prünster I (2013) Poisson driven stationary Markovian models. Technical report
4. Mena RH (2003) Stationary models using latent structures. PhD Dissertation, University of Bath
5. Mena RH, Ruggiero M, Walker SG (2011) Geometric stick-breaking processes for continuous time Bayesian nonparametric modeling. J Stat Plan Infer 141:3217–3230
6. Mena RH, Walker SG (2005) Stationary autoregressive models via a Bayesian nonparametric approach. J Time Ser Anal 26:789–805
7. Mena RH, Walker SG (2007a) On the stationary version of the generalized hyperbolic ARCH model. Ann Inst Stat Math 59:325–348
8. Mena RH, Walker SG (2007b) Stationary Mixture Transition Distribution (MTD) models via predictive distributions. J Stat Plan Infer 137:3103–3112
9. Mena RH, Walker, SG (2009) On a construction of Markov models in continuous time. METRON Int J Stat LXVII:303–323
10. Pitt MK, Chatfield C, Walker SG (2002) Constructing first order autoregressive models via latent processes. Scand J Statist 29:657–663

Chapter 37
Claim Sizes in the Compound Poisson Process from a Bayesian Viewpoint

Gamze Özel

Abstract The typical assumption of independence among claim size distributions is not always satisfied in risk modelling. In this study, the exchangeable claim sizes are considered aggregated claims that are obtained via compound Poisson process. Exchangeability of the claim size is obtained by the conditional independence, using parametric and nonparametric measures for the conditioning distribution. A Bayesian analysis of the proposed model is illustrated with Turkish Earthquake Insurance Claims Data between 2000 and 2003.

37.1 Introduction

In risk modelling, when the number of claims follows a Poisson process, the aggregated claims amount can be modelled by the CPP which is denoted by $\{X_t, t \geq 0\}$ and defined as follows:

$$X_t = \sum_{i=1}^{N_t} Y_i \qquad (1)$$

where $\{N_t, t \geq 0\}$ is a homogeneous Poisson process with parameter $\lambda > 0$ and Y_i, $i = 1, 2, \ldots$ are independent and identically distributed random variables independent of N_t. In risk theory, $\{N_t, t \geq 0\}$ is the number of claims performed to the company during the time interval $(0, t)$ and Y_i, $i = 1, 2, \ldots$ is the ith claim size. Hence, X_t is the aggregated claims up to time t [5, 6].

Bayesian methods are very useful in actuarial science since it yields to learn about the whole distributions of quantities rather than just obtain an expected value

---

G. Özel (✉)
Department of Statistics, Hacettepe University, Ankara, Turkey
e-mail: gamzeozl@hacettepe.edu.tr

for each parameter. These methods allow to include many levels of randomness in the analysis through the use of prior distributions for each parameter, which highlights the uncertainty regarding individual distributions or parameters [1]. In addition, the posterior distribution can be updated and obtained when new information becomes available. The existing literature on the Bayesian analysis of CPP includes, but is not limited to, works by [7, 8] and [2].

In this study, the CPP is used for the aggregate claim and a variety of loss distributions to model the size of individual claims. Bayesian methodology is used to fit distributions to the claim sizes. This approach assumes that all parameters in the distribution are variables. The results derived by using Bayesian methodology with those from classical statistics are compared to see which had the best fit with the data.

37.2 Estimation of the Claim Count and Claim Amount Distribution

In classical risk theory, it is very common to assume a homogeneous Poisson process for the claim arrival process since this assumption simplifies the derivation of the total claim amount distribution [3]. Also, a gamma distribution model is frequently assumed to describe the usual right-skewed shape of the claim size distributions. In this section, the Bayesian estimation of the total claim count and the total claim amount in a future time period are given [4]. A Bayesian estimation of the CPP is obtained by calculating the posterior means of their cumulative distributions, also called their predictive cumulative distribution functions. In order to set ideas, started with CPP model with independent claim sizes, $Y_{ij} \sim f(y|\theta)$ for $i = 1, 2, \ldots, N_{tj}$. Due to the independence of the Poisson processes N_{tj} and the claim sizes Y_{ij}, inference for λ and θ is done separately. The number of events N_{tj} follows a homogeneous Poisson process independent of the expenditures sizes Y_{ij} which are gamma distributed with parameters a and b. We then start by assuming that X_{tj} is a CPP with independent claims, that is, Y_{ij} are all independent for $i = 1, \ldots, n_j$ and $j = 1, \ldots, m$. If the prior knowledge on (a, b) can be represented by $\pi(a) = Ga(a|\alpha_a, \beta_a)$ and $\pi(b) = Ga(b|\alpha_b, \beta_b)$ independently, then the posterior distribution, given the sample, is characterized by the conditional distributions.

An illustration based on Turkish Earthquake Insurance Claims Data between 2000 and 2003 is given. Posterior inference for the insurance data is carried out by implementing Gibbs samplers for the CPP model. For sampling from each of the conditional distributions random walk Metropolis-Hastings steps are used. This proposal distribution is centered at the previous value of the chain and has a variation coefficient of one. The Gibbs sampler was run for $100,000$ iterations with a burn-in of $200,000$ keeping every $10th$ observation after burn in to reduce the autocorrelation in the chain. Finally, a predictive analysis for the aggregated expenditures of a patient in a year is carried out. For that the information about the frequencies of occurrence of the claims is needed, modelled by the Poisson processes and in particular by the intensity λ. Considering that we observed $1,342$ claims in the year, then the posterior distribution for the intensity of the claims per year, λ, is $Ga(1341; 10.658)$. Therefore, the posterior mean rate is 12.42 claims per person per year. With the posterior distribution for λ and the posterior predictive distribution for the whole sequence of claims, we obtain the posterior predictive distribution for the aggregated claims in a year for one individual.

References

1. Bernardo JM, Smith AFM (1994) Bayesian theory. Wiley, Chichester
2. Dudley C (2006) Bayesian analysis of an aggregate claim model using various loss distributions. Dissertation thesis for Master of Science in Actuarial Management. Heriot-Watt University, Edinburgh
3. Hogg RV, Klugman SA (1984) Loss distributions. Wiley, Toronto
4. Makov UE, Smith, AFM, Liu, Y (1996) Bayesian methods in actuarial science. Statistician 45(4):503–515
5. Özel G (2011) On certain properties of a class of bivariate compound Poisson distributions and an application to earthquake data. Revista Colombiana de Estadística 34(3):545–566
6. Özel G, Inal C (2012) On the probability function of the first exit time for generalized Poisson processes. Pakistan J Stat 28(1):27–40
7. Pai JS (1997) Bayesian analysis of compound loss distributions. J Econom 79:129–146
8. Panjer HH, Willmot GE (1983) Compound Poisson models in actuarial risk theory. J Econom 23:63–76

Chapter 38
Land Rental Market and Agricultural Production Efficiency: A Bayesian Perspective

Haoran Yang

Abstract In this study we used a generalized true random effect (GTRE) stochastic frontier model to estimate the impact of land rental market on agricultural production efficiency in rural Chongqing, China. After we provide a complete description of land institutions and land rental market environment in the study area, we employed Bayesian methods of inference in estimation to investigate this impact. In practice we used Metropolis-within-Gibbs sampler to do the inference because it is straightforward to impose regularity conditions on production function implied by economic theory in this process. We also used cross entropy (CE) to facilitate our choice of candidate-generating density. Empirical results showed that the land rental market has the potential to transfer land from less efficient farm household to more efficient farm household, but hardly has impact on the management level of farm household who participates in the market.

38.1 Introduction

The share of agricultural sector in China's GDP is declining over the last thirty years, while people who used to live on agriculture didn't decrease proportionally, which is one of the reasons of persistent rural-urban inequality in China.

Economic theory suggests that a functioning land rental market can improve agricultural production efficiency and raise income of farmers, whereas land rental market in developing countries like China never functions properly. This leaves us the questions if land rental market can promote agricultural productivity in China and which factors may affect the functioning of land rental market. The impact of land rental market on agricultural production efficiency is twofold: one is that

H. Yang (✉)
Center for Development Research(ZEF), Walter-Flex-Str.3, Bonn, Germany
e-mail: yhaoran@uni-bonn.de

if land rental market can transfer land from less efficient farm households (LFHs) to more efficient farm households (MFHs); another is that if land rental market can improve technical efficiency of a typical farm household. These are the two research questions we are going to investigate in this study.

38.2 Research Methodology

We used stochastic frontier model to analyze the impact of land rental market on agricultural productivity. The GTRE model can incorporate time-varying technical efficiency and persistent technical efficiency in a systematical way [3].

We use Bayesian statistics in estimation, because it is straightforward to impose curvature constraints suggested by economic theory [4] and it holds good property for small sample size [5] in Bayesian procedure. The GTRE model has the form

$$y_{it} = f(x_{it}; \beta) + \alpha_i + v_i + u_{it} - z_i$$

where y_{it} is the output of farm i ($i = 1, \ldots, I$) at year t ($t = 1, \ldots, T$), x_{it} denotes the input vector, β is the parameter vector of production function which we assume as a multivariate normal distribution, α_i represents farm-specific effect which is time persistent, v_{it} denotes stochastic error of production, both α_i and v_{it} distributed normally, v_{it} is time-varying technical efficiency, and z_i is time persistent inefficiency term, both v_{it} and $z_i t$ distributed exponentially.

Economic theory suggests that the production function $f(x_{it}; \beta)$ should be a monotonic, concave, and homogeneous of degree one in x_{it} [2]. We employ the translog production function in empirical study which has the form

$$\ln y_{it} = \alpha_i + \sum_{j=1}^{N} \beta_i \ln x_{it} + \frac{1}{2} \sum_{j=1}^{N} \sum_{k=1}^{N} \beta_{jk} \ln x_{jk} \ln x_{jk} + \gamma_t t + \gamma_{tt} t^2 + v_i + u_{it} - z_i \quad (1)$$

Homogeneity of degree one in inputs implies that $\sum_{j=1}^{N} \beta_i = 1$ and $\sum_{j=1}^{N} \beta_{jk} = 0$. Based on these two conditions homogeneity can be imposed explicitly by normalization. $f(x_{it}; \beta)$ is monotonic in x_{it} implies the first partial derivative of $f(x_{it}; \beta)$ on the element of x_{it} is nonnegative; $f(x_{it}; \beta)$ is concave in x_{it} if and only if the Hessian matrix H of $f(x_{it}; \beta)$ is negative semi-definite. We can answer the first question by comparing the willingness to pay for additional unit of land (i.e., the value of marginal product of land) between LFHs and MFHs. If the willingness to pay for MFHs is higher than LFHs, a competitive land rental market can transfer land from LFHs to MFHs because MFHs can bid for the land at higher rent. To answer the second question, we can estimate the impact of participation in land rental market on technical efficiency of production of farm households. This means we are trying to explain $E(u_{it}) + E(z_{it})$ by a set of explanatory variables in which participation in land rental market is the interested variable.

Denote $\theta_{it} = E(u_{it}) + E(z_{it})$; the corresponding technical efficiency level is given by $TE_{it} = exp(-\theta_{it})$. The efficiency function can be elaborated as

$$TE_{it} = \delta_{1i} + \delta_1 H_{it} + \delta_2 P_{it} + \delta_3 AGE_{it} + \delta_4 M_{it} + \delta_5 F_{it} + \delta_6 D_m + \omega_{it}$$

where H_{it} is Herfindahl index which measures crop diversity, P_{it} is the proportion of net rented land in total operational land area, F_{it} is the number of plots which measures land fragmentation, and AGE_{it} denotes age of household's head. $M_{it} = FM_{it}/(FM_{it} + AP_{it})$, where FM_{it} denotes expenditure on farm machine and AP_{it} denotes expenditure on animal power. D_m is village dummy variable which measures village effect on technical efficiency, $\omega_{it} \sim N(0, \sigma_\omega^2)$. We assume the constant terms $\delta_{1i}(i = 1, \ldots, I)$ have the same normal distribution; the distribution of slope parameters of efficiency function has a multivariate normal form.

38.3 Data and Estimation Strategy

The data we used in this study is from the fixed observation point of Research Center of Rural Economy of Ministry of Agriculture of China.

Metropolis-within-Gibbs sampler was used in posterior inference of (1). Except for the parameter vector β, all other parameter posterior conditional densities have the standard form, exhibiting either Gamma distribution or normal distribution. The posterior conditional density of β is given as follows:

$$\beta|h, u, z, \gamma, \eta \sim N(\overline{\beta}, \overline{V})1(\beta \in M) \tag{2}$$

where $\overline{V} = (h \sum_{i=1}^{I} X_i' X_i)^{-1}$, $\overline{\beta} = (\sum_{i=1}^{I} X_i' X_i)^{-1}(\sum_{i=1}^{I} X_i'[y_i + u_i + z_i \iota_T])$, and $1(\beta \in M)$ is the indicator function which equals to one if β satisfy the monotonic and concave conditions and zero otherwise.

We use the independent China Metropolis and Hastings algorithm to generate random draw from (2). To implement this procedure, we need to find an appropriate candidate-generating density. Here CE can be employed to do this job. In the first step, we estimate the stochastic frontier without constraints (without the indicator function) by using Gibbs sampler. We get the posterior density function for β in which parameter vectors that satisfy monotonic and concave conditions can be regarded as "rare event." In this case CE method can be used to explore this "rare event" and formulate a proper candidate-generating density [1]. Then we reestimate the function's parameters in the constraint model by using Metropolis-within-Gibbs sampler in which inference of posterior conditional density of β in (2) is based on independence chain Metropolis-Hastings algorithm using the candidate-generating density we obtained in step one. After we draw a random parameter vector from candidate-generating density we calculate an acceptance probability and accept this draw randomly.

A hierarchical Bayesian model can be set up for the efficiency function, and Gibbs sampler can be used for posterior inference.

38.4 Empirical Results

We calculated the value of marginal product of land (shadow price of land). Then we group farm households into three categories according their efficiency score at mean: LFHs, moderate efficient farm households, and MFHs. The results of ANOVA show that the shadow price of land for MFHs is significantly higher than LFHs, which imply that in a competitive land rental market, land will be transferred to more efficient farm households and land-use efficiency will be improved. But the market transactions we observed are not consistent with the prediction of theory, because farm households which rent in land are significantly different from farm households in the MFH group.

Participating in land rental market can improve farm management level of farm household by adjusting their farm size from marketplace, but the gain is negligible, based on the results of efficiency function. These results may suggest that participation in land rental market can improve technical efficiency of agricultural production either by enhancing land-use efficiency or by promoting farm management level, but the land rental market environment which is characterized by high transaction cost, informative asymmetry, and opportunism of landlord can't provide enough incentive for farmers to fully use the potential of land rental market.

38.5 Conclusions

Land rental market indeed has the potential to improve agricultural production efficiency in the research area. But the potential of land rental market is not fully realized because of incomplete market environment.

CE method can be used to formulate candidate-generating density in Metropolis-within-Gibbs algorithm and improve posterior inference efficiency.

References

1. de Boer PT, Kroese DP, Mannor S, Rubinstein RY (2005) A tutorial on the cross-Entropy method. Ann Oper Res 134:19–67
2. Chambers RG (1988) Applied production analysis: a dual approach. Cambridge University Press, Cambridge
3. Colombi R, Martini G, Vittadini G (2011) A stochastic frontier model with short-run and long-run inefficiency random effects. Department of economics and technology management working paper, University of Bergamo, Italy. Available via DIALOG. http://hdl.handle.net/10446/842/ofsubordinatedocument.Cited10Jan2013
4. O'Donnell CJ, Coelli TJ (2005) A bayesian approach to imposing curvature on distance functions. J Econom 126:493–523
5. Tsionas EG, Kumbhakar SC (2012) Firm heterogeneity, persistent and transient technical inefficiency: a generalized true Random-effect model. J Appl Econom 25:1–23

Part VI
Suggestions for Young Researchers

Chapter 39
The Point Is…to Publish?

Fulvia Mecatti

Abstract Writing papers is an essential part of the research process. Researchers have a professional obligation of disseminating their results, making them available for others to use to enhance common scientific knowledge. Besides the fun of sharing their own ideas and views, to publish is essential in order to actually have a scientific career. Although scientific writing certainly has its own conventions and standards, I suspect there is no a unique true recipe making the trick. As a matter of fact I do not have any. However my quite long time in the academic arena has given me a pretty clear idea about how I do and do not like things done. In this paper I will be giving my personal view and rules, in the hope that sharing my own experience would do some good to others as it did for me.

39.1 From the Big Technical Rules to *My* WHW Rules

Shortly after being asked to give this talk I realized I said yes a bit too quickly and started having second thoughts. Of course I was grateful since addressing young statisticians is certainly a part of my job that I do love. I thought I could give them a good nice recipe for disseminating effectively the results of their research.

Except, I did not have any. I needed preparation.

So I started the easy way, surfing the Internet and in fact finding a huge amount of material: big technical rules *"Do this and that," "Go this way not that," "The right thing to do is...is not,"* and on and on along these lines. Sadly, they also contradict slightly too often. For instance somewhere in the process I came across this really convincing statement:

Make sure to do all the research *before* start writing

F. Mecatti (✉)
Department of Sociology and Social Research, University of Milan-Bicocca Building U7 Via Bicocca degli Arcimboldi, 8, 20126 Milan, Italy
e-mail: fulvia.mecatti@unimib.it

and a couple of Web pages after that I found myself totally disoriented in front of the following assertion:

> Start writing early and often. Don't leave all the writing to the end and then try and write it all out in one go.

Trying not to be discouraged I went on and I was really starting to believe in this

> Pack all the good ideas in the first pages of your manuscript, even into the abstract: readers are busy people and get bored quickly

only to stumble shortly after upon this

> Disclose ideas slowly and leave the best results until the last: you will capture reader's attention from the beginning to the end

Should I choose the one to be trusted best and feed that to my young audience as the *truth*? Would you go and try to decide which one is *the correct way* to do things? I would not and in fact I did not. Thus, no big help over the Internet, except perhaps a raising suspect that there might not be such a thing as the big technical rules. Besides I knew for a fact that I was about to face much better net-surfers than I am. They were so not needing me that I decided to leave that to them.

Second in my list were books.

So I went to the library and in a couple of my favorite bookshops: shelves after shelves after shelves of confident books about *what to do and do not*, and *how to make it work*, in your mother language or in a second language. Tons and tons of literature, there already. They could go and help themselves, they did not need me either. In the end I made my decision: I would have left all that to them and gone personal. I would have made it clear that they were not going to be given some tight big truly rules about how to do or not to do things, for I was not sure I could do that or even it would exist. Instead I was pretty sure I could tell them how *I like* or *I do not like* things done, hoping *my* experience would do some good to them too as it did for me.

Thus it will be *my view* and *my rules* and only occasionally a few basic technical rules, so basic it would be impossible not to agree with. And in this case it will be no more than a couple of them. Mostly they will be *Why, How, and Where*, namely WHW rules. I will be talking about that in different contexts and declinations of the subject of disseminating the results of scientific research, statistical research indeed.

39.2 The Point *Is* to Publish

In my opinion, there is little point in having good ideas if you cannot communicate them to others. Besides teaching, being a researcher is the fun of a scientific/academic career. In the everyday life this translates being always in (at least) one of the following three statuses:

1. You have an article *nearly* finished.
2. You have an article *about* to be started (or two, three, etc., no actual limit here).
3. You are stuck somewhere in the *middle* of an article.

As a matter of fact writing articles is an essential part of the research process. But it is not the end. There is also the dark side, those gory details leading to the *publish or perish* paradigm. I am sure you know what I am talking about here, both despite of and due to your young academic age. As scientific authors you do write for the fun and pure joy of communicating ideas. However you also need to write for tenure and promotions. In order to actually have a scientific career, you do need to be published. When the goal is to go through the steps of an academic career you are essentially what you have published.

So the point *is* to publish and it is time for some WHW rules, assuming that something to publish already does exist. That is, we are not talking about the research process, instead we have already some nice research results worth disseminating. Thus *why* to publish? To publish is the main way to communicate your ideas and at the same time it is functional to your academic career. It is essential to be aware about that and to find soon your own balance between these two sides of the very same moon. Besides, research is a shared matter and it has its own ethic. Researchers have a professional obligation to both perform research and disseminate the results of that research, as objectively and as accurately as possible.

> Science is not an individual experience. It is shared knowledge based on a common understanding of some aspect of the physical or social world. During the birth of modern science in the latter half of the 17th century, many scientists sought to keep their work secret so that others could not claim it as their own. Prominent figures of the time, including Isaac Newton, often avoided announcing their discoveries for fear that someone else would claim priority. The solution to the problem of making new discoveries available to others while assuring their authors credit was worked out by the secretary of the Royal Society of London, Henry Oldenburg. He won over scientists by guaranteeing both rapid publication in the society's Philosophical Transactions and the official support of the society if the author's priority was questioned. Oldenburg also pioneered the practice of sending submitted manuscripts to experts who could judge their quality. Out of these arrangements emerged both the modern scientific journals and the practice of peer review. Once results are published, they can be freely used by others to extend knowledge. But until the results are so widely known and familiar that they have become common knowledge, people who use them are obliged to recognize the discoverer by means of citations. In this way, researchers are rewarded by the recognition of their peers for making results public. [1]

Although digital technologies are creating new forms of publication, publication in a peer-reviewed journal remains the most important way of disseminating a complete set of research results. The importance of publication accounts also for the fact that the first to publish a view or a finding—not necessarily the first to discover it—tends to get most of the credit for the discovery.

39.3 How to Share and the RELUKE Rule

The leading tool for disseminating research results is a paper. A paper is an *organized* description of the research—from conjectures and hypotheses to conclusions—intended to *instruct* the reader. A statistical paper is on average 8–15 pages long. Of course there are also thesis and textbook. However, in statistical research, book writing is rare. It is mostly subsequent to a long research, summarizing and perhaps concluding a wide research project. This is also the case of a final thesis, each of us has at least a couple of experiences about. Let us keep it for another occasion since they are quite a different story; let us focus over papers and the key ways to share a paper. There are two main ways for academic authors to disseminate a paper

1. By presenting it as a talk or as a poster at a conference or at a seminar
2. By publication in a peer-reviewed journal

In Stats—where almost all my experience is—these are certainly the most frequent medium and also the most rated for both the bright high goal of disseminating scientific ideas and the dark dirty goal of making a living and getting promoted in academia.

How to do it is the major challenge.

A way to start that makes sense to me is to think and state what my objectives are, what I want from what I am doing, whether writing a paper for publishing or preparing a presentation for a talk. First I do not want to waste everybody's time: the author's, audience's, referees', and readers'. And second is the RELUKE rule. That is the re(ad), l(isten), u(nderstand), and (lu)ke rule: I want people to listen/read me till the end of my presentation/paper. I want people to understand what I am saying/wrote. And I want people to like it. Presenting or writing it is of course about you but not you alone. As already mentioned, a paper is a description of the research *and* is intended to *instruct* the reader. Never forget about the audience. Whatever you are doing for your research dissemination, whether preparing a talk or a manuscript, you are talking or writing *for* someone. The process is a constant in and out: it is about you the author, your results, and what you want to share (*in*); at the same time it is about your audience/readers, someone *else* who is going to listen/read you (*out*). Whatever the way you chose to share, it essentially means: know the focus of your paper *and* be clear of your intended audience, working out what you believe they already know and what they might not know. As a conclusion I would like to share another good rule I was given when I was a young researcher myself:

> If you did want to learn how to write: *read a lot*.
>
> If you did want to learn how to prepare and give a talk: go to conferences, attend seminars and *listen a lot*.

A golden rule indeed.

39.4 Presenting a Paper in a Talk: *My* WHW Rules

Why presenting a paper in a talk? Many journal articles do begin as talks presented in professional meeting such as a conference, a workshop, and a seminar. Public presentation is often the first step toward writing an article and trying to publish it. It is a good way to present intermediate results of a larger research. It certainly is a valuable opportunity for getting criticism and suggestion so that to refocus your research. It is an occasion for networking and meeting other people interested in the topic and researching in the same field. For me, presenting intermediate results at a conference helps in meeting deadlines and imposing discipline to my work.

How? A presentation with slides is the usual option. The challenge here is how to make it work, according to the RELUKE rule. You do have this precious thing of an audience; what you want is to make them listen to your talk, understand it, and like it. Assuming you have nice results to present, first of all: *time*. You are always allotted a well-defined time for your presentation, which you are going to meet out of respect for your audience, for the speakers scheduled after you, and for the organizers. Conference slots are usually no longer than 15 min; it might be 10, rarely 20. For a seminar the time is longer but still constrained, usually 45 min or 1 h. Thus you cannot put *everything* into your presentation, into your slides, into your talk. Maybe you are so lucky that your audience includes the guru(s) very experts of your subject. Except possibly them, no one else will have the time to do the mental process that costed you so long to produce your results. Be aware of your time and careful in selecting what you can give people in no more than those minutes. Secondly, *speak up* and discuss spontaneously your own slides. In Stats, reading a prepared talk is a rare practice. It might seem comforting, you may think to be better in control, but you will just result as boring. Not worth the effort! Third, control your speech *speed*. Speak slowly and loud, making your words clear, especially if English is your second language. Avoid sounding rushed or breathless. If you realize you are not going to make it in saying all you would have wanted to say, cut something from your talk. Less yet clear is far better than more yet incomprehensible. A good deal of rehearsal will do the trick here.

Where? You may be giving your talk at a seminar or at a conference. To give a seminar you are normally invited by the organizer, usually at a different institution from your own. In choosing a conference you should consider a basic classification. A conference may be *general* as for a periodic meeting of a statistical society; some examples are the biannual scientific meeting of the Italian Statistical Society every even year; the annual Joint Statistical Meeting of the American Statistical Association, every August joint with 6 other societies; the biannual meeting of the International Statistical Institute every odd year. Usually you do not need to be a member of the society to attend the meeting; however if you are going to present a paper it can be requested for a membership of at least one of the authors. Any statistical paper, concerning any fields of statistical research, submitted to the

Table 39.1 Top ten tips for a really hideous presentation

1.	Present slides very dense and in text size no greater than 10
2.	Pack your slides with many horrible formulae
3.	Abound in different text fonts and colors
4.	Use many acronyms and technicality that only you (and possibly some very experts in the field) can understand
5.	Take your cell calls in the middle of your speech
6.	Avoid originality and personality
7.	Read your slides instead of a spontaneous discussion
8.	Speak in a low voice and make a lot of ehm-ehmming
9.	Speak unclear and very quick
10.	Pass your slides fast and furious

scientific committee of the conference, will be considered for the presentation. Otherwise a conference may be *specific*, meaning focused upon a special subject such as this very conference primarily intended for young Bayesians. In this case, before submission you need to be checking carefully if the subject of your paper meets the areas of interest for that specific conference. A conference, either general or specific, is usually organized in sessions that we can classify in 4 main categories:

1. Keynote or plenary session, where a senior speaker is invited for a lecture on a topic he/she is supposed to be an expert.
2. Invited or specialized session, where an organizer is in charge to think to a special topic, usually innovative or where research is very active, and to collect a small group of invited speakers who have published on that topic. It is common practice to add a discussant to the session, who is another speaker not presenting his/her own paper but in charge to start and animate the discussion with questions and comments over the papers presented in that session.
    The remaining two types of sessions are probably more interesting and more likely for young statisticians.
3. Contributed session, which is a set of spontaneously submitted papers, sometimes gathered around a common topic but not necessarily.
4. Poster session, which is alternative to an oral presentation of your paper. You are allowed a space where to post your paper, displayed as a poster of fixed size, and you are requested to guard your own poster for all the session and be available to interested people for face-to-face discussion.

There is great variability in the review-acceptance-refusal process of a conference and each conference committee usually has its own rules. However any committee will state very tight deadlines for submission of title, abstract, short paper, long paper, and revised versions.

39.5 Writing a Manuscript for Submission: *My* WHW Rules

With a written paper, people has of course more time to read as compared with listening a 15 min talk. Sadly this not necessarily implies to pay attention. Readers are busy people. Assuming you have something to write about, writing a manuscript needs per se special care and skills. And it always takes far more time than planned.

Why? To think that your main objective is *to publish* is in fact a limited thinking. For me, the worst thing that could happen to my manuscript is not rejection. It would be to get published and then lay there unread and uncited. Still the reluke rule! We are all very busy (and sometimes slightly arrogant). As a consequence to pay time and attention to a paper not strictly related to our current research is a tough yet quick decision. I myself have the habit to decide to go on and read a paper on the basis of title, abstract, and author(s). This essentially means that if you are not yet an affirmed author—and you need to work hard and long to become one of them, presenting and writing a lot—chances are that you will get to be read on the basis of title and abstract. My suggestion is to put quite a lot of cure on those. Still, even if I decide to actually give the paper a go and read through it, I would start with the beginning and the end. That is, I would read the Introduction and the Conclusions. After that I normally would go through the entire paper as long as I was clear it would be potentially useful to the very research I am working on or I am interested in at that particular time (and in that case chances are I will be needing to read and reread several times and always spend over it more time than initially planned). My point here is you do put a lot of care into the Introduction and the Conclusions also. These are the parts I myself usually find most difficult to write.

How? Statistical scientific writing of course has its own formal conventions about article writing. We will be articulating this a bit but very quickly. Because the good news is if you accomplished the golden rule *read a lot* you will be absorbing subliminally such rules and conventions. Papers are mostly organized the same way, the key composing parts being summarized in Table 39.2. The Title is a challenge: it should be short but clearly telling the reader what your article is about. The Title and Keywords are fundamental for having your paper properly indexed and showing up in computer searches. A simulation is intended as an exercise carried on in a wholly controlled environment, usually a large set of Monte Carlo runs over artificially

Table 39.2 Standard components of a statistical paper

Title
Abstract and Keywords
Introduction
Notation
Method and Theoretical Results
Simulation and Application
Conclusions
Acknowledge
References
Appendix

Table 39.3 Top ten tips for writing truly boring dull papers [3]

1.	Avoid focus
2.	Avoid originality and personality
3.	Make the article really really long
4.	Do not indicate any potential implications, applications, and developments
5.	Leave out illustrations (too much effort to draw a sensible drawing)
6.	Omit necessary steps of reasoning
7.	Use abbreviations and technical terms that only you and some specialists in the field can understand
8.	Make it sound unnecessary serious with no significant discussion
9.	Focus only on data
10.	Quote numerous references for trivial statements

generated data. On the other hand an Application is performed by means of a *real* data set. Application results might be the main objective of the whole research as well as the way to show that the proposed method is actually applicable and able to produce an outcome of use. Both Simulation and Application are intended to offer clear empirical evidence to your claim and many journals rate both crucial for a manuscript to be considered for publication. Conclusions are meant to help the reader reorganize ideas. The Reference, as well as citation in text, must be double-checked: you do not want to give the reader ground to doubt your reliability. Finally your manuscript might need an Appendix for boring proofs and tricky analytical details.

As for some tips on writing well, I have just one: keep it simple *and* specific. Readers—especially the intelligent ones—have higher probability to be impressed by ideas and clear expressions rather than by elaborate constructions and excess of words. Rather say it once clearly, than several times verbosely. Vague and murky statistical writing is perfect ground for just one suspect: reduced to simple it does make little sense or it is too banal to be worth saying. On the other hand, simple does not mean trivial or superficial. You do not need to be wordily in order to be deep and accurate. Simple and specific serves both the author and the reader: it saves reader's time *and* author's reputation. Here are my personal rules to implement the simple and specific writing:

1. Avoid superlatives and go easy with adjectives: *"excellent", "the best and the most", "very useful"*, and the like.
2. Be careful with adverbs and judging: *"the right thing to do", "obviously", "of course", "rather", "certainly"*, and so on. Here the question is: *who for*?

You do not need much of that to be precise and meaningful. You can delete most of that with no sacrifice to meaning and sounding. My first drafts are usually heavily judgmental, stuffed of superlatives, and of *adding* things such as adjectives and adverbs. In revising I simply cut it away and very rarely I have finally found my paper missing any of them. If English is not your mother language—as it is not for me—my main advice is: go and get a good spell and grammar checker, either in your editing software or over the Internet. Moreover, I have two preferred suggestions [2]

1. Prefer active verb over an excess of nominalization. For example, in

    We conducted a review of the evolution of the method

    review *and* evolution *and* method is an overload of nominalization. Reexpressing as *active verb* the statement is clearer, isn't it?

    We reviewed how the method evolved;

2. Prefer active verb (if available) over a generic verb like *to do, to get*, and the like. For instance in

    Modeling of the random-effects distribution nonparametrically has not yet been done

    done is a generic, empty verb. How about rewording as

    The random-effects distribution has not been modeled nonparametrically.

Enough, right? Also ... don't you think that try and write keeping in mind all those *how to, avoiding, do and do not* would distract you from writing the ideas and results? Surely it does for me. So my hint is: first have a first draft done. Write for writing, just put down all the ideas and results, trying not to worry over the form, the good language, the embellishment. Only then, you do remember all the *good writing tips* and start revising your first draft. This basically means a lot of rereading and reworking your paper. For me this is the stage where time is at maximum risk of spinning out of control. Be careful and do not be carried away by obsessive revising: perfection is not to be reached in this world. And of course I have developed my own two tricks to stop obsessing over the rereading and reworking of a manuscript:

1. Be patient and leave the draft there for a while. Go for a walk, distract yourself with stuff. After some time, a couple of days or possibly a week, you go back to your manuscript for another reading with fresh attention. You will be amazed about how packed of overlooked details and typos to fix and rearranging your draft still needs. Of course this should not stop the submission forever, my suggestion here is: once it *is* enough. If you are lucky enough of not being a compulsive reviewer, be patient as well: try and not to submit the very second after writing the last word.
2. Look for a second pair of eyes (even a third could do). Ask one or two of your colleagues, advisors, or seniors to read your manuscript before submission. Be prepared for criticism and accept it graciously. If you bothered to ask, you should bother to listen. If you were just looking for a pat on your back or uncritical encouragement, shouldn't you have asked your pet?

39.6 ... and the Grand Final: Submission and the Publishing Process

We tend to think that having the paper written and submitted is the heavy part of the process. This is only the lucky case. Scientific publishing is a long and interconnected team effort, involving the journal, the author(s), and the reviewers.

How to choose the journal where to try and publish that very paper, with its peculiarity? A big name, an excellent reputation, mainstream, and a high impact factor are indisputable classic criteria. They usually do but not always. Maybe you need to fly lower, having a nice manuscript which is not first-class statistics and still worth publishing. For me the average time for that journal from submission till acceptance and publication is an important selection factor. This is why in my own reviewing activity I always do my best to meet deadlines. The acceptance/rejection rate of the journal, when known, is my second important selection factor. In case you have no idea where to start, look into your own paper and consider journals you are citing at your turn. Go navigating its Web page and check the journal editorial policy and board. Have a look at the latest issue, it is often available online; otherwise go to the library. In my experience I occasionally considered to start from the top, submitting to a major journal, in the hope of trying and getting a reputable high-level review, even if along with almost sure rejection. Nevertheless it helped both in revising the manuscript and targeting the journal where resubmit. First choices, even the most carefully made, happen to be wrong: be prepared to submit more than once, to more than one journal. However contemporary multiple submission is usually a big sin, which is either not tolerated or forbidden by most journals. Except for some e-journals that clearly declare permission of multi-submission and multi-publishing, you will be honestly submitting to one journal at a time. You will not be risking that the very same manuscript will be sent to the same reviewer from two different journals. That would damage unnecessarily your reputation.

And *what* will they do with your submission? After submitting a paper, be prepared to wait. Your submission has triggered a delicate interconnected serial process. First the editor in chief will be passing the manuscript to his/her associate editor with competence on that specific subject. Then two referees will be asked to peer-review the manuscript and will be given adequate time to fit this further commitment into their agenda, normally no less than 5 weeks. The peer-review process is usually but not always double-blinded: the referees are always anonymous to the author, not necessarily the other way round. The reviewers will motivate their own recommendation in a report to the associate editor which is in charge of the final decision about your manuscript. A third reviewer might enter the picture to disentangle two conflicting recommendations. Only at this point the submitting author is notified the final decision, normally alongside a report of comments and suggestions. It thus might take a while for the entire process to conclude. Wait patiently, but not forever. After a three/four months time of silent waiting you should write a polite email to the editor asking when you may expect a response. The dreaded-looked forward final decision will be one and only one of three:

1. Rejected—often with a polite encouragement to resubmit
2. Accepted conditionally on major/substantial revision
3. Accepted with minor revision—the lucky case

A fourth option

> Accepted as it is, no revision required

is a non-impossible event with probability close to zero. It can happen.
As a matter of fact, reviews tend to come with a lot of disappointment

> they did so NOT get it!

and much Charlie Brown thinking

> nobody understands me ... nobody appreciates my work ...

going back and forth between the two with different and personal intensity. We love statistics and we are passionate authors, that is, human and very Italian. Anyway, sooner or later you gotta get over it. So first take a good breath and some days to overcome the shock. Be sure to be over the emotional though natural part of the process before going back to rework your paper. That is, be sure you are just working on the paper (not making them see or making them be sorry). Talking as a reviewer myself, after all they *did* an effort. Peer review is a totally volunteer activity. Reviewers are experts, but human beings, they *can* be wrong (and sometimes even unnecessarily nasty). However, in most cases, they just did their job looking for weak points, obscure methodologies, and unconvincing statements, offering constructive criticism even when suggesting rejection, giving advice for moving a manuscript from the *unacceptable* class to the *publishable* status. If a reviewer misunderstood a point, that point probably needs to be made clearer. In every review (even in the wrong and gratuitously rude ones) there *is* some good for your manuscript and your research. The whole point of the refereeing saga is: it *is* a service and it is there for you to use. Find out of the reviewers' job how you can make your manuscript stronger and acceptable for publication. Look for the good in the review and use it.

When considering resubmission, revision should be *always* the case, whether or not you are going to choose a different journal after a rejection. In my experience, trying to submit to another journal without any revision normally does not change the final result, ending in another rejection and very similar reviewers' comments. Of course I know that for a fact ... for having tried myself. Also, you cannot exclude that the second journal would send your manuscript to the very same reviewers as the first journal did. And this would be very unrespectful of the reviewer job, no matter he/she was right or wrong. If you are resubmitting upon a conditional acceptance, my suggestion is to accompany the revised manuscript with a clear report about the revision you made; according to another WHW rule, you should be reporting about *What*, *HW*, and *Why* you did or did not in reworking the manuscript.

This may or may not include a point-by-point answer to the reviewers. After resubmission, some more waiting. But now you know what to expect. It will be the editor's decision to ask the reviewers if they were totally satisfied and if you had been addressing all the comments. The report just mentioned will be useful in that phase.

So, it might require quite a while and a lot of work, still the saga generally has its happy end: the editor congratulating you and asking for proofreading your galley proof.

Acknowledgements I would like to thank Francesca Ieva and Anna Maria Paganoni for inviting me to give this paper. I had a great time twice: first addressing young attendees at BAYSM2013 and then writing this paper. I accepted because they asked so nicely I could not say no. Now I know what the fun of becoming a senior might be.

References

1. National Academy Press (1995) On being a scientist. National Academy Press http://www.nap.edu/openbook.php?record_id=4917&page=R1
2. Little RJ, Wilson S. WRITE Statistics RIGHT! Tips on good writing style for R&M researchers. Bureau of the Census http://sitemaker.umich.edu/rlittle/files/writestatsrev.pdf
3. Sand-Jensen K (2011) How to write consistently boring scientific literature. Oikos 116:723–727. doi: 10.1111/j.2007.0030-1299.15674.x

Printed by Publishers' Graphics LLC
LMO131213.15.16.86